IN VIVO

The Cultural Mediations of Biomedical Science

PHILLIP THURTLE and ROBERT MITCHELL, Series Editors

IN VIVO

The Cultural Mediations of Biomedical Science

is dedicated to the interdisciplinary study of the medical and life sciences, with a focus on the scientific and cultural practices used to process data, model knowledge, and communicate about biomedical science. Through historical, artistic, media, social, and literary analysis, books in the series seek to understand and explain the key conceptual issues that animate and inform biomedical developments.

THE TRANSPARENT BODY
A Cultural Analysis of Medical Imaging
by JOSÉ VAN DIJCK

GENERATING BODIES AND GENDERED SELVES
The Rhetoric of Reproduction in Early Modern England
by EVE KELLER

THE EMERGENCE OF GENETIC RATIONALITY
Space, Time, and Information in American Biological Science, 1870–1920
by PHILLIP THURTLE

BITS OF LIFE
Feminism at the Intersections of Media, Bioscience, and Technology
Edited by ANNEKE SMELIK and NINA LYKKE

Bits of Life

■ *Feminism at the Intersections of Media, Bioscience, and Technology*

Edited by

ANNEKE SMELIK

and

NINA LYKKE

UNIVERSITY OF WASHINGTON PRESS ▪ *Seattle and London*

This book was made possible by the financial support
of the Netherlands Organization of Scientific Research.

Printed in the United States of America
15 14 13 12 11 10 09 08 6 5 4 3 2 1

University of Washington Press
P.O. Box 50096, Seattle, WA 98145, U.S.A.
www.washington.edu/uwpress

The interview with Donna Haraway (chap. 3) was
previously published in full in the Danish journal
Kvinder and in *The Haraway Reader* (Routledge, 2004).

Library of Congress Cataloging-in-Publication Data
Bits of life : feminism at the intersections of media, bio-
science, and technology / edited by Anneke Smelik and
Nina Lykke.
p. cm. — (In vivo)
Includes bibliographical references and index.
ISBN 978-0-295-98809-2 (pbk. : alk. paper)
1. Feminist theory. 2. Human reproductive technol-
ogy. 3. Information technology. I. Smelik, Anneke.
II. Lykke, Nina.
HQ1190.B573 2008
305.4201—dc22 2007048825

The paper used in this publication is acid-free and
recycled from 20 percent post-consumer and at least
50 percent pre-consumer waste. It meets the minimum
requirements of American National Standard for Information Sciences—Permanence of Paper for Printed
Library Materials, ANSI Z39.48–1984.

Contents

Acknowledgments

Bits of Life evolved from a series of seminars and conferences that took place between 1999 and 2005 within an international exchange program titled "Media, Cultural Studies, and Gender: Looking for the Missing Links," which was generously funded by the Netherlands Organization of Scientific Research. We thank Rosi Braidotti for her inspiring leadership and Trude Oorschot for her efficient management of the program. Over the course of the program, the following universities were involved: in the Netherlands, the University of Utrecht and Radboud University Nijmegen; in Denmark, the University of Southern Denmark and the University of Aarhus; in the United Kingdom, the University of Lancaster; and in the United States, the University of California, Santa Barbara. Three of the annual conferences were co-organized and co-funded by the Danish research project "Cyborgs and Cyberspace: Between Narration and Sociotechnical Reality," under the direction of Nina Lykke, with funding from the Danish Research Agency's FREJA (Female Researchers in Joint Action) program (1999–2003). Substantial contributions were also made by the Women's Studies Department of the University of Lancaster. In addition, Linköping University, Sweden, and the Swedish Research Council supported publication of the volume.

Final thanks go to Sandra Janssen, Teun Dubbelman, and Jasper Sluijs for their careful preparation of the manuscript.

A. S. and N. L.

Bits of Life

An Introduction

ANNEKE SMELIK and NINA LYKKE

The title *Bits of Life* invokes a figuration that signifies today's cultural fusion of the biological and the technological. It also points to the current proliferation of different discourses on "life," indicating that there are many "bits" of life that we need to think through. This is a daunting task, given that life is so ancient and discourses on life are so changeable. In this book, the figuration "bits of life" is interpellated in order to allow us a critical distance from currently emerging discourses and practices that make "life" into an object of scientific, religious, and cultural attention and fascination. Discourses on life are, after all, entangled in gendered, racialized, and sexualized truth regimes, claiming to represent the final word of science or religion. These are saturated with power relations, which are structured in a complex, weblike manner that differs significantly from classical binary oppositions between the haves and the have nots.

The phrase "bits of life" means that we do not engage with "life" as a whole but with its many manifestations in art and popular culture, the humanities, and the sciences. It forces us to strike alliances across academic disciplines as well as across the "two cultures" of the arts and the sciences. The emphasis on life in contemporary culture also brings in the return of discourses and practices about the human body. In *Bits of Life*, we explore and evaluate the current reinvestment in the human body, a body that is full of life as well as disease and death, reconfigured by technology and bombarded by bits and bytes of information and experience.

Since World War II, the biological and the technological have fused and amalgamated in new ways. The third culture that, in the late 1950s, Snow (1993) envisaged for the future was closer than he perhaps imagined. In 1960, the notion of the cyborg (cybernetic organism) was coined as part of the evolution of the science of cybernetics and early space-flight research, predicting for the near future a radical redesign of bodies, which would make humans and animals fit for life in outer space

(Clynes and Kline 1995). The cyborg figure that later was remade as a prominent feminist figuration (Haraway 1991a, 1991b), suggested a material and semiotic dissolution of the boundaries between organism and machine. With the typical techno-optimist rhetoric of the 1960s, the famous science fiction writer Arthur C. Clarke (1964), one of the first to introduce the cyborg figure to a broader public, predicted that cyborgs would soon make the distinction between organism and technology obsolete. Deleuze and Guattari (1972, 1980) philosophically elaborated the point that there is no longer a clear distinction between ourselves and our technological environments. Instead, there are complicity, intimacy, and promiscuity between the given and the acquired, according to the "process ontology" introduced by the poststructuralist generation (Braidotti 2006).

The entanglement of the biological and the technological, of "man and machine," is in itself not new, but the sheer expansion and all-pervasiveness of that entanglement is quite staggering, as is the ever-accelerating speed at which the two have been merging in the last decades. The technological redesign and reconfiguration of bodies and environments becomes more and more a part of everyday life. Against this background, a rethinking of bodies as well as environments becomes a pressing issue for information science and the biological sciences. As life bits, whether carbon- or silicon-based, are transformed, the body threatens to fall apart into "components," to decompose down to its molecular structures, which can be reassembled in new and unexpected ways and remediated in endlessly changing shapes. The human body can no longer be figured either as a bounded entity or as a naturally given and distinct part of an unquestioned whole that is itself conceived as the "environment." The boundaries between bodies and their components are being blurred, together with those between bodies and larger ecosystems.

Moreover, related convergences, which also contribute to the blurring of boundaries between the biological and the technological, are forcefully being put on the agenda today. For example, the discourses of the info- and biosciences are becoming more markedly connected. Keller (2000: 127) notes that "the conceptual traffic between engineering and biological sciences . . . never [has] been heavier or more profitable." Other scholars argue that it is not possible or desirable to separate media, science, and technology. Kember (2005) writes that "new media forms are . . . increasingly in-formed by biology." The significance of biotechnology vis-à-vis media is underscored by Haraway (2000: 26), who is convinced that biology will supersede film or literature as one of the great "representing machines" of this century.

Our cultural "practices of looking" (Sturken and Cartwright 2001) are changing our perceptions of bodies, technologies, and ourselves. As Bolter and Grusin

(2000) argue, the logic of endless remediations of new multimedia creates a paradox of immediacy (that is, of the medium appearing to be a transparent window on the world) and hypermediacy (that is, of the medium foregrounding its form, breaking the illusion of immediate access to the real). This phenomenon contributes significantly to the blurring of boundaries between bodies and technologies. Mediated by a technologically enhanced gaze, bodily micro- and macroworlds, from cells to planets, are becoming an unquestioned part of our everyday life outlook.

The figure "bits of life" is grounded in this convergence between biogenetic sciences and regimes of visualization, pointing to the "emergent paradigm of biocultures."[1] The relation between biology and culture is not harmonious but conflictual. Our figuration assumes a posthuman definition of the body: a body that is not-one. While taking the fragmentation of the body-in-ruins from postmodernism, this collection of essays moves beyond the particular postmodernist economy of affect that has wavered between euphoria and nostalgia. *Bits of Life* assumes a more sober grounding of the questions surrounding the biocultures of today, which do not refer back either to humanist bodily integrity or to the anthropocentric assumption that human bodies are the only ones that matter. The centrality of "life itself" (Rose 2001) opens up the perennial discussion of how bios relates to culture, and the contributors to this volume address the different facets as well as the extent of the shift in relations between bioscience and culture today. Some pay more attention to the pole of bios; others interrogate the cultural realm.

Because technoscience and bioculture constitute a site of anxiety, in this book we sketch some conflictual or contradictory contours of biocultures, tracking the multiple power relations that circulate in their midst. It is quite striking how often certain boundary transgressions or new possibilities opened up by biotechnology are shut down or blocked. For example, in chapter 5, Amade M'charek and Grietje Keller show how new configurations of three-person "parenthood," enacted by in vitro fertilization (IVF) technologies, are erased by hegemonic notions of coupled heterosexual parenthood, and Celia Roberts, in chapter 4, argues that the interdependent ecosystems in which hormones are now seen to flow as "global fluids" are met by cultural messages about individualized consumers' responsibilities for preventing hormonal "exposure." The cultural imaginary also responds with anxious resistance to technoscientific transformations. In chapter 7, for example, Jackie Stacey analyzes the way in which "technologies of imitation," realized with new cloning techniques, are cinematically translated into the sexualized dangers of the masquerading female monster. And in chapter 9, Anneke Smelik reveals a similar cultural imaginary, which turns the technocultural matrix of cyberspace into what is, once again, the familiar story of the flight from the body and the fear of and fascination with the (culturally denied) maternal womb.

If bioculture is indeed something new, it nevertheless does seem to have features that are all too familiar. From the repetition of those old stories we may even deduce certain cultural blockages that are uneasily coding the implications of bio/technoscientific changes. The question is whether cultural power relations necessarily change when bodies, memories, genetic identities, sperm and eggs, and hormones can be experienced and technodesigned in different ways. In these conservative times, we may tend to focus on the hegemonic cultural practices that try to contain the emergent possibilities of bio/infotechnologies. Yet, as José van Dijck (chapter 8) and Jenny Sundén (chapter 10) show, digital technologies also open up new stories and social changes. And Karen Barad (chapter 11) and Rosi Braidotti (chapter 12) opt for a more dynamic approach to understanding life as a mode of becoming.

In this book, then, the figuration "bits of life" is meant to be an evocative term that can help us map changes and transformations and strike a middle road between the metaphorical and the material. By invoking the figure "bits of life" against this background of a diversity of blurred boundaries and machinic assemblages, we hope to present an adequate and condensed feminist analysis of present-day technoscience and biocultures. *Bits of Life* captures a certain technophilic sensibility as one of the distinctive traits of feminist studies of media, biocultures, and technoscience. As we have seen, the body in its many material-semiotic modes and codes is (still) an important pivot. Feminist studies of digital media and information technology have criticized the utopian vision of a flight from the body. Similarly, critics of the new reproductive technologies and genetics have expressed concern about the potential enhancement of bodies, such as it is envisioned by some geneticists and reproductive scientists. This book rests on the assumption that these concerns need not result in technophobic rejections and appeals to "pure" nature and "uncontaminated" bodies. Rather, in exploring current reconfigurations and remediations of bodies and embodied subjects as "bits of life," we pursue a technophilic yet critical approach as well as thoroughly reflected articulations of new ethical standards. This critical approach, in highlighting the problems as well as the potentials of biocultures and technoscience, constitutes a shared frame of reference for the contributions to this volume.

In *Bits of Life*, we bring together feminist studies of media, biocultures, and technoscience. So far, each of these aspects of feminist studies has developed independently, and there has been too little cross-referencing. We break with the tradition in feminist scholarship that looks at biotechnologies, information technologies, and media as separate phenomena; instead, we create synergy and bridge the gap between the different studies of biocultures, new media, and technoscience. We do so by building on long-standing traditions in feminist cultural studies of technoscience while also looking for the convergences between different approaches.

To understand the new modes and codes of our increasingly technologized lives, we need new forms of media literacy as well as new tools and frameworks for interpreting technobodies. The rapidly changing information and communication technologies and the fast-growing repertoire of biotechnologies have produced the need for new communicative and analytical protocols. These in turn necessitate increased forms of familiarity with the technologies themselves. The convergences between and among digital media, information technologies, and biotechnologies call for specific methodological tools and theoretical approaches. The framework of blurred boundaries (human/machine, nature/culture, technology/organism, sex/gender) heralded by the cyborg figuration has proved a fruitful one for feminist cultural studies of technoscience, and it constitutes a point of departure for this book. But, as we have already argued, we want to push the discussion still further and enlist the "bits of life" figure in our questioning of fixed entities and boundaries.

This implies, first, that the practice of science criticism and the philosophy of science require an interrogation from many perspectives—from ontology, epistemology, ethics, and politics, for example. A more hybrid and dynamic approach is needed, one that goes beyond the science wars of the 1990s (Gross and Levitt 1994), which itself is an echo of the "two cultures" debate of the 1950s (Snow 1993), as Maureen McNeil shows in chapter 2. A new alliance is needed among feminist theory, cultural studies, and studies in science and technology. This volume aims to further such an alliance.

Second, the return of the "real body," in all its thick materiality, spells the end of the linguistic turn, in its postmodernist overemphasis on textuality. With this book, we hope to effectuate a change in feminist cultural studies, urging the field to go beyond classical notions of semiotics and hermeneutics, and on to explorations of new material-semiotic approaches that can lead to a "materialized deconstruction that literary Derrideans might envy" (Haraway 1997: 102). We reintroduce a certain materialism and realism in exploring new forms of a cinematic or digital aesthetic that moves beyond representation. The new biotechnological discourses bring to the fore the material foundations of the embodied self, including its biological and genetic material. The emphasis on life marks a shift away from the deconstruction of layers of textuality, and toward an understanding of the inextricable entanglement of material, biocultural, and symbolic forces in the making and unmaking of the subject.

Third, this methodological shift poses a challenge to the dominating social constructivist trends in feminist theory. The emergence of life itself as a subject of investigation highlights the limits of social constructivism as a method of accounting for the "hybrid" structure of contemporary technological culture. A phenomenon like the emergent biocultures cannot be dealt with in the conventional language and methodology of the social sciences. It is a transversal phenomenon that calls into question a cluster of factors, and of multiple effects. Digitization and globalization add their impacts to these complex processes. Therefore, we must develop scientific thinking at the intersection of different domains and learn to think in terms of processes and interrelations. The emergence of "bits of life" as a subject forces a new relationship between the natural sciences and the social sciences, restructuring the position of the embodied and embedded material foundations of "life" as well as the social and symbolic representations that sustain them.

Fourth, our approach raises the question of feminist activities in academia. Here, we can briefly offer some background information about the joint project of which this book is one result. *Bits of Life* grew out of a series of European seminars and conferences, funded primarily by the Netherlands Organization of Scientific Research but also by the Danish Research Agency, that took place within an international exchange program at the beginning of the new millennium. Over the years in which these seminars and conferences were held, academics from the Netherlands, Denmark, Sweden, England, and the United States were involved in a particular program titled "Media, Cultural Studies and Gender: Looking for the Missing Links." The project was also linked to the Advanced Thematic Network in Activities in Women's Studies in Europe (ATHENA), a European network of departments and programs of women's studies from more than a hundred European universities.[2] Thus this project of contemporary feminist theory has been funded by national governments as well as by the European Union.

Europe is a transnational entity, and so our work is situated in both multiculturalism and polylingualism (Griffin and Braidotti 2002). Much of our work has been translated into English from other languages. Our terms of cultural reference are necessarily different from those in the North American context, and so we may choose different cultural metaphors or implicitly express our theoretical alliance with the materialism of continental philosophy. Moreover, the European context of our work means that we may have different experiences of power relations in biocultures and technoscience. For example, stem cell research and genetic engineering are a part of our university environments. The politics of life and death also vary considerably, from the U.S. right-wing crusade against abortion to the law allowing euthanasia in the Netherlands. As a result, many of the authors of this book engage with U.S. scholarship and U.S. popular culture as partial outsiders. For the authors of *Bits of Life*, such differences along cultural and national lines have led to lively discussions and thought-provoking exchanges, which we hope have found their way into this collection.

Bits of Life is divided into four parts, each part consisting of essays that are published here for the first time. The book also includes an interview with Donna Haraway (chapter 3), one of the most prominent scholars in the field of feminist technoscience studies.

Part 1, "Histories and Genealogies" (chapters 1–3), presents technoscience studies as a shared theoretical and methodological frame of reference for the book and introduces the reader to the complex history of the interlocking and overlapping fields of feminist studies, cultural studies, and science and technology studies. Thus the first two chapters map out genealogies of the hybrid field of feminist cultural studies of technoscience.

In chapter 1, Nina Lykke sets herself the impossible task of mapping an implosion—the dynamic, open-ended implosion of interdisciplinarity. She draws up a diagram of the three fields of overlap that constitute feminist cultural studies of technoscience: feminist studies, cultural studies, and science and technology studies. In so doing, she spells out the key dynamics of each field as well as the many intersections between and among them. Lykke argues that the founding act of each field consists of a deconstruction. In the case of feminist studies, this is the deconstruction of gender; in cultural studies, that of the opposition between high and low culture; and in science and technology studies, that of the positivist notion of science as rational progress. She concludes the chapter by invoking the figure "bits of life" as a follow-up to the figure of the cyborg, blurring the boundaries between organism and technology, and between matter and discourse. "Bits of life," then, is an imploded knot from which an infinite number of threads can be untangled.

In chapter 2, Maureen McNeil carefully traces the various histories of this interdisciplinary area, investigating feminist contributions from such disciplines as cultural anthropology, literary studies, art history, film studies, and science fiction studies. She discusses the complicated relationship between the interdisciplinary field of British cultural studies and the field of technoscience studies. She explores the reasons why it took a long time for mainstream British cultural studies to become interested in technoscience issues, and why feminists took the lead. McNeil also emphasizes the strong contributions of feminists and other cultural studies scholars to the reorientation of science and technology studies from a focus on the high culture of science to the meaning of technoscience in everyday life.

In chapter 3, the interview by Nina Lykke, Randi Markussen, and Finn Olesen with Donna Haraway, the interview subject comments extensively on her style as an author, moving in between the literary and the theoretical. Surprisingly for an atheist, Haraway links her unorthodox style to the Catholic tradition of "unnameableness." She makes clear that the area of language, narrative, and rhetoric cannot be divorced from science without the risk of serious reductionism. For her, the use of categories—the effort to rename and resituate them for tactical reasons—is an act of modest witnessing. She also comments on the position of feminists as both insiders and outsiders in science and technology studies.

Part 2, "Reconfigured Bodies" (chapters 4–7), looks at different kinds of bits of life that are known for their ability to reconfigure the biological body in the age of biocultures and technoscience. We are talking here about such bits of life as hormones, eggs and sperm, germ cells, and genes. These are analyzed through the different lenses of scientific discourses, public debates, in vitro fertilization, science documentaries, and Hollywood cinema.

In chapter 4, Celia Roberts explores shifting discourses in the field of endocrinology that have been central to the production of scientific models of sexual difference. Endocrinology is one of the fields where a shift from a "mechanical" to an "informational" body took place as early as in the beginning of the twentieth century. The understanding of hormones as "messengers of sex" is an example of the convergence between scientific discourses and info- and biosciences. Against this background, Roberts analyzes the implications of the discursive shift from a bounded, hormonal body to an unbounded, fluid body extending beyond the sealed boundary of the individual body. She pursues these issues by comparing scientific discourses and public debates on hormone-replacement therapy and on endocrine-disrupting chemicals in the environment.

Chapter 5, by Amade M'charek and Grietje Keller, shifts the perspective from hormones to germ cells, from endocrinology to new reproductive technologies. M'charek and Keller perform ethnographic research to analyze how reproductive technologies both disturb and reconstruct notions of sex, gender, parenthood, kinship, and nationhood. To make their point, the authors look at convergences between two very different uses of IVF techniques, one in human infertility treatment and one in cattle breeding. By comparing the performance of IVF technologies in the human and animal realms, they are able to show how categories like kinship and parenthood are constructed by technological interventions. In the age of genetics, IVF technologies reconfigure the relations between genetic and social parenthood. An understanding of the way in which IVF technologies travel between the contexts of human and animal reproduction may help to undermine the biological determinist idea that genetic parenthood and the social couple should coincide.

In chapter 6, Mette Bryld and Nina Lykke take on the Swedish science photographer Lennart Nilsson, who has become world-famous for his images of the fetus in the womb. Bryld and Lykke compare the Swedish and U.S. versions of one of Nilsson's most recent films on human reproduction. While both films are firmly rooted in a positivist and objective version of natural science, the Swedish film has a much more conservative framework. It adheres more strongly to the objective tone of the genre of the documentary, thus mystifying heterosexuality and reproduction, while the U.S. version includes a personalized story that allows for the more experiential narrative of a multicultural couple. Moreover, the scientific story of reproduction is significantly different in the two films, with the U.S. version favoring a more "modern," equality-oriented tone.

In chapter 7, Jackie Stacey explores reconfigured bodies in fictional cinema, analyzing the way in which popular Hollywood cinema stages genetics. How does cinema work around the dilemma that the gene has no visual signifier? Stacey illustrates the potentials for genetic engineering of (new) bodies in relation to gender and sexuality by discussing in detail two science fiction films, *Gattaca* and *Species*. Both foreground cultural anxieties about complex and confusing relationships between and among cultural identities, genetic makeup, and genetic engineering of bodies. The juxtaposition of the two films enables Stacey to discuss the quests for detection and deception that are part and parcel of the genetic imaginary. She shows how *Gattaca* puts a masculine desire for mastery of technologies on display, while *Species* exposes genetically engineered femininity as a monstrous threat to scientific control.

In Part 3, "Remediated Bodies" (chapter 8–10), the focus shifts from reconfigurations of bodily "life bits" to remediation—that is, to bodies, body parts, and embodied subjectivities (re)produced by the movement from one medium to another. In focus here are the digitization of personal memories, tunnel images in science fiction films and biomedical documentaries, and body assemblages in a hypertext novel.

José van Dijck, in chapter 8, explores the way in which personal memory is affected by the possibilities of digital storage, tracing the cultural fantasy of a universal memory machine as a desire for total recall and total control. Van Dijck takes the example of a digital project, "MyLifeBits," which allows the user to store personal bits of life in the computer. While she finds that digitization of memory gives rise to new forms of materiality in practices of remembering, she remains critical of the promises of new software programs to fix and preserve private memories. She calls for a deeper understanding of the social changes that come along with new technologies for remediating what is dearest to us—our very personal memories.

In chapter 9, Anneke Smelik explores the digital remediations of the human body in science fiction films and biomedical documentaries. She compares a recurring moment in both genres: the spectacular ride through a tunnel, either into cyberspace or into the human body. Discussing examples that range from B movies like *Freejack* to the sophisticated trilogy of *The Matrix*, she points out the heavy traffic between and among virtual/real, inner/outer, and fact/fiction that is performed by the tunnel ride. Smelik suggests that kinetic representations of cyberspace are informed by images of the inner space of "real" bodies, as taken from documentaries (her example is the prime BBC series *The Human Body*). She argues that "inside out" and "outside in" are collapsed into an imaginary space that thoroughly confuses the real and the virtual.

Jenny Sundén, in chapter 10, also explores the virtual. She reads a hypertext novel, Shelley Jackson's *Patchwork Girl*, for its questioning of the limits of bodies, and of life itself. This digital novel, intensely involved with issues of monstrosity and femininity, tells the story of the female mate of Frankenstein's monster. Working from the materiality of hypertext fiction, Sundén shows how information technology comes to act as reproductive technology that reproduces not only texts and images but also the life itself of the she-monster. *Patchwork Girl* has been read as a postmodern celebration of a poetics of the fragment, but Sundén argues that it actually speaks of being fragmented as severely painful. "Bits of life" become the quintessence of the monster's fractured subjectivity.

Part 4 (chapters 11 and 12) moves on to philosophy, presenting different approaches to posthuman materialism. In chapter 11, Karen Barad takes her lessons from Schrödinger's cat, that famous rhetorical device for paradoxical exposure of the materiality of quantum mechanics. She tries to displace the natural sciences' claims to objectivity, arguing instead for an understanding of the mutual implications of the material and the discursive. Life, she says, is not an inherent property of individuals but is performed through its material phenomena and discursive practices. Therefore, the matter of life can be understood only in its dynamic process of becoming.

The notion of becoming is also central to chapter 12, by Rosi Braidotti, who proposes an understanding of life as "zoe," by which she means a vitalistic and generative life. Zoe allows for an affirmative appreciation of life. Braidotti extensively traces the history of "life as zoe" in philosophy and goes on to develop the concept of a sustainable self. She embraces the ethical principle of affirmation by putting forward what she calls a "sustainable nomadic ethics." The concept of nomadism points to an understanding of the self and of the subject as a dynamic process of continuous becoming. In order for that nomadic self to be sustainable, Braidotti argues, the subject needs endurance. Endurance, for her, is joyful affirmation as the inherently positive potential of the subject. Thus Braidotti develops a new figuration of living subjectivities in the posthumanist mode.

Bits of Life highlights the search for tools and theories by which it becomes possible to analyze the complex interplay among textual, visual, imaginary, technological, and biological dimensions of bodies, of subjects, and of life. We hope that the reader will find various ways of assessing methodological and theoretical frameworks for feminist research of media, biocultures, and technoscience. Each of the chapters that follow tackles "bits of life" in their many manifestations in art and popular culture, the humanities, and the sciences.

NOTES

1. We take this phrase from the Biocultures Project of the University of Illinois at Chicago, which organized a conference and webcast in March 2005 titled "Biocultures: An Emerging Paradigm."

2. See Braidotti, Nieboer, and Hirs (2002); for related material, see also www.let.uu.nl/womens_studies/athena/outcomes.html (retrieved March 2, 2007).

PART 1

■ *Histories and Genealogies*

1 ▪ Feminist Cultural Studies of Technoscience

Portrait of an Implosion

NINA LYKKE

This chapter gives an introductory overview of feminist cultural studies of technoscience, the hybrid and interdisciplinary field that makes up a shared frame of reference for the contributions to this book.[1] I present here some interdisciplinary key dynamics of the field, to make things easier for readers of *Bits of Life* who are not familiar with the ways in which feminism, cultural studies, and technoscience studies—that is, the central components of feminist cultural studies of technoscience—have clashed as well as merged in recent decades.

The productive dynamism of the field builds very much on a relentless interdisciplinary and transdisciplinary openness to inappropriate and impossible connections, and to explorations of theoretical in-between spaces where it is possible to think differently. To portray such dynamism is not an easy task. The risk is of freezing it into quasi-stable assemblages of bounded entities or building blocks, which easily may take on a tinge of "essential truth" about "fixed components." To avoid such counterproductive freezing, and to fulfill the didactic purpose of giving an overview, I borrow two methodological devices from Donna Haraway.

First, I apply Haraway's notion of the "imploded object" (Haraway 1997: 12; see also chapter 3, this volume). I analyze feminist cultural studies of technoscience as a knot into which different strands of the interdisciplinary nodes of research interests are imploding in an open-ended process. Second, I use a Venn diagram as an analytical engine to spell out the "components" of feminist cultural studies of technoscience.[2] In this chapter, however, the Venn diagram is used in a playful mood, one inspired by the way in which Haraway once ironically interpellated the famous semiotic square of classic structuralism as a "clackety structuralist meaning-making machine" (Haraway 1992: 304). I let the Venn diagram act like the semiotic square in Haraway's text, where, in its noisy insistence on a machinelike, modern way of acting, it fulfilled an ironic poststructuralist demand for exposure of the technologies of research design. In an analogous vein, the Venn diagram can insist noisily on its machinic presence in my text, thus exposing the subjective and situated moment of design and construction.

FROM SCANDINAVIA WITH LOVE: A POSITIONING

Here, as a prelude to my overview of feminist cultural studies of technoscience, I will briefly contrast my Scandinavian genealogies with the Anglo-American ones that are traced in detail by Maureen McNeil in chapter 2. In so doing, I underline the complexities of the field and stress its relentless resistance to canons and master narratives.

The label "cultural studies of technoscience" was introduced into the Anglo-American context in the early 1990s. Before that time, as McNeil stresses, people were doing this kind of research, but without the joint nodal point that, for better or for worse, is engendered by a naming device. As a Scandinavian scholar of feminist cultural studies of technoscience, I recognize the genealogies outlined by McNeil as important international trends that have inspired me and my Scandinavian colleagues in the field. From a genealogical point of view, however, our story is somewhat different. The most significant difference is perhaps the fact that studies of science and technology have been explicitly performed from the perspective of the humanities at several Scandinavian universities since the early 1980s. Feminist endeavors that were defined as feminist cultural studies of science and technology were part of this trend. When I started conducting this kind of research in Denmark, in the mid-1980s, international feminist science and technology studies formed a significant platform. But Scandinavian sources of inspiration were equally important. For example, the department for interdisciplinary studies of technology and social change at Linköping University, in Sweden, has been vital. From its start, in the late 1970s, the department stressed the importance of interdisciplinary cultural studies of technology. In the early 1980s, it hosted several research projects on cultural studies of technoscience, such as one titled "The Machine and the Humanist," and its first Ph.D. degree was earned with a dissertation on Strindberg and machines (Kylhammar 1985). A pioneering feminist research project titled "Women's Culture, Men's Culture, and the Culture of Technology: Looking for a Border-Crossing Language" was also being carried out at the department in the early 1980s. The latter project led to an important dissertation on the sewing machine (Waldén 1990).

The feminist technoculture studies that came out of this department at Linköping University became a crucial inspiration for me and other feminists at the University of Southern Denmark. In 1985, we started a program in gender, culture and technology studies; it still exists, and over the years it has fostered and sustained a number of bigger and smaller research projects, with an outspoken profile in feminist cultural studies of technocience (Lykke and Braidotti 1996; Bryld and Lykke 2000). The program also generated a project titled "Cyborgs and Cyberspace: Between Narration and Sociotechnical Reality" (1999–2003), which, together with the Dutch-funded "Media, Cultural Studies, and Gender: Looking for the Missing Links," sustained part of the networking that led to the current volume. Other examples of the early institutionalization of cultural studies of technoscience in Scandinavia are programs at the Danish universities of Aarhus and Aalborg in the 1980s, which integrated a feminist approach. In 1985, for example, the Aalborg group hosted a major Nordic conference titled "Women, Natural Sciences, and Technology."

Against the background of the early institutionalization of interdisciplinary cultural studies of technology in Scandinavia, the historical moment of introducing and establishing the label "cultural studies of technoscience" on the international scene, in the early 1990s, signified the confirmation of already labeled and explicitly promoted research programs. I will not go more deeply into these genealogical considerations and the ways in which a Scandinavian perspective, as well as broader continental European perspectives, might be able to sustain more comprehensive feminist understandings of the geopolitically varied types of knowledge production that have contributed to the transnational emergence of feminist cultural studies of technoscience. The more limited and modest aim of this chapter is, as mentioned, to give a brief, didactic overview, with a focus on interdisciplinary key dynamics.

TO MAP AN IMPLOSION

How to perform the impossible task of mapping an implosion? In physics, an implosion is defined as an explosion directed inward instead of outward. If the cathode-ray tube of your television implodes, you will see only a small white dot on the screen; the major part of the process will be invisible, displaying none of the spectacle of the explosion. Likewise where the phonetic meaning of the word "implosion" is concerned: in phonetics, the distinctive feature of an implosive consonant is that only the closing of the mouth is heard; the rest is silence charged with meaning. Thus an implosion, while invisible, silent, and unspectacular, is nevertheless a very dynamic process. I think it was precisely the double edge of dynamism and invisibility that prompted Haraway to use the imploded knot as a methodological tool. This double edge makes the imploded knot an apt tool for the metaphorical articulation of how apparently frozen objects of study and "self-evident" entities emerge from dynamic transformation processes.

Haraway opens up seemingly fixed entities, such as "the end-of-millennium seed, chip, database, bomb, fetus, race, brain, and ecosystem" (Haraway 1997: 12), precisely by considering them as implosions. She describes them as "offspring of implosions of subjects and objects and of the natural and the artificial" (ibid.). According to Haraway, such imploded objects or knots are devices that can engage the researcher or analyst in an infinite process of untangling, where new relations continue to emerge. To grasp this methodological point in another way, see the video (Paper Tiger Television 1987) in which Haraway, using a ball of yarn from which she continues to pull out new threads, visually illustrates the process of infinite methodological untangling of imploded knots, and the article (Haraway 1994) in which she uses the game of cat's cradle as a methodological device to outline the intersectional and ever-changing relationships between and among feminism, cultural studies, and science and technology studies. Feminist cultural studies of technoscience can be aptly described as an implosion—or as a ball of yarn or a game of cat's cradle—in this open-ended sense. Here, I will try to indicate the implosive dynamics by portraying these fields as a Venn diagram, to be understood in an ironic, poststructuralist mode, where, as in Haraway's semiotic square (Haraway 1992: 304), the diagram's modernist, essentializing taxonomic ambitions are exposed to ridicule through its enginelike, utterly mechanical way of creating distinctions that seem to be always already on the verge of breaking down, in kaleidoscopic fashion, and of giving way to new divisions, themselves only momentarily stable.

If the Venn diagram functions as an engine, let me now fill it with semantic fuel. Figure 1.1 consists of three intersecting circles, which visually represent feminist cultural studies of technoscience as a moment of implosion of feminist studies[3] (top), cultural studies (left), and science and technology studies, or STS (right).

In keeping with the open-endedness and playful mood with which the Venn diagram is used here, the main circles—feminist studies, cultural studies, and science and technology studies—should be understood not as entities but as interdisciplinary nodes of research interests that have had strong community-building potentials. Each has led to periodically recurring conferences as well as to the emergence of journals, scholarly networks, and associations. In other words, they have all acted as scientific communities, according to mainstream definitions. But they can also

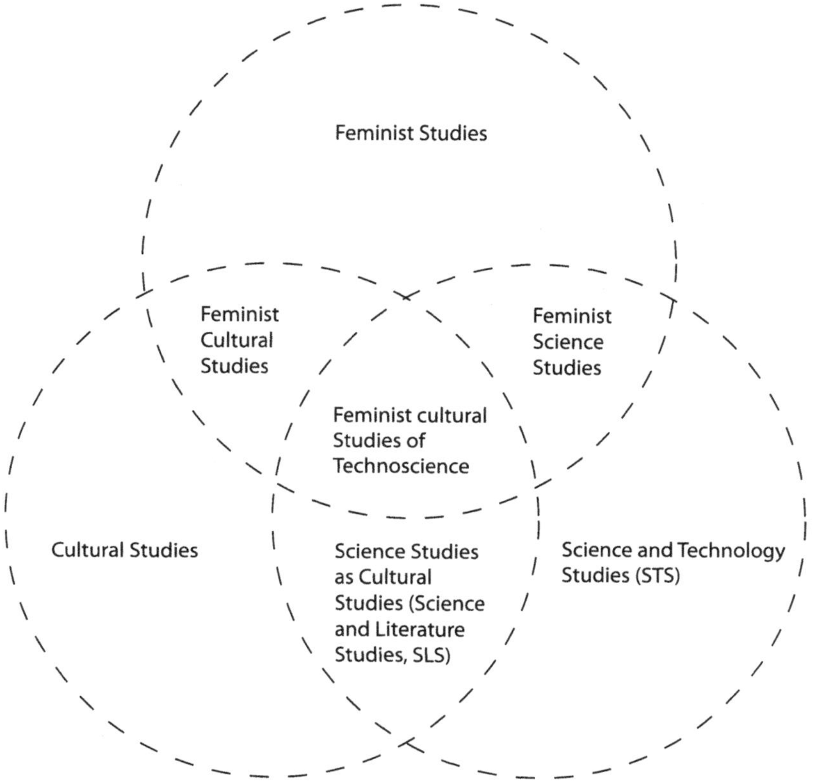

Fig. 1.1. Venn diagram representing feminist cultural studies of technoscience

be viewed as fluid sites in constant intersection and interaction, and as nodes of open-ended research interests that have produced continuous tensions and synergies. To underline the open-endedness, the circles are drawn with dotted circumferences, to emphasize the external permeability and mutual internal intersectionality of the imploding areas.

In the next section of this chapter, I will spell out some of the key dynamics implied in figure 1.1. I will focus on interactions in the four fields of overlap. First, however, let me emphasize that my presentation involves certain genealogical considerations. As we know, genealogies are contested areas. Who and what is included or excluded in the constructed histories of academic areas? What kinds of narrative structures are erected in order to highlight these particular histories? What kinds of "unsettled relations" (Thornham 2000) can be detected between different narratives about the same field of study? Such questions are especially crucial as well as controversial when we consider genealogies and key dynamics of open and heterogeneous areas, such as the ones we are dealing with here. Space permits me to mark only some of the specific dynamics of each field, without doing justice to their details and the differences within them. Therefore, I will focus on two key dynamics for each overlap space in figure 1.1, thus drawing a map of feminist cultural studies of technoscience as a hybrid area, one that productively amalgamates and transforms many different research trends and interests.

FEMINIST CULTURAL STUDIES: INTERACTING KEY DYNAMICS OF FEMINIST STUDIES AND CULTURAL STUDIES

The field of feminist cultural studies has close connections with the interdisciplinary field of cultural studies. One of the founding acts of cultural studies was the critique of the high-low culture divide (Storey 1993). What we understand to be culture is no longer exclusively defined by elite culture, which linked the concept of culture to a narrow canon and to essentialist ideas about relations of class, nationhood, gender, ethnicity, and cultural superiority. Instead, with Raymond Williams (1988), we now understand culture to be practices of everyday life, comprising all kinds of popular culture and media.

Feminist cultural studies cannot adequately be described as just another chapter in the overall history of cultural studies. This point has been emphasized by feminist chroniclers of the relationships between feminism and cultural studies (Franklin, Lury, and Stacey 1991; Radway 1998; Thornham 2000). There are many parallels and overlaps between feminist studies and cultural studies—for example, the political commitment to cultures and subjectivities of "inappropriate/d others" and marginalized groups (Haraway 1992; Minh-ha 1986/87). Nevertheless, tensions, "unsettled relations" (Thornham 2000), and "lack of overlap" (Franklin, Lury, and Stacey 1991) have also been part of the picture, for several reasons. First, women's perspectives were initially ignored in the field of cultural studies, which started out with a focus on workers and youth cultures, from an implicitly masculine outlook. The classic volume *Women Take Issue* (Women's Studies Group 1978) testifies to the struggles and theoretical rethinking that were necessary in order for feminism to be introduced into cultural studies in the 1970s. Second, a principal strand of feminist studies focused on cultural issues right from the start, independently of the emerging cultural studies community. Because the women's movement of the 1970s wanted to change the cultural relations of everyday life, a "theory of gender oppression in culture" (de Lauretis 1990: 267) had a prominent place in early feminist theories. Third, when we shift the perspective from a narrative of cultural studies to one of feminist cultural studies, a somewhat different set of themes calls attention to itself. One of these themes is the body as a controversial and important issue (Thornham 2000). And with the body, the role of science in culture is put on the agenda as well (Franklin, Lury, and Stacey 1991).

FEMINIST SCIENCE STUDIES: INTERACTING KEY DYNAMICS OF FEMINIST STUDIES AND SCIENCE AND TECHNOLOGY STUDIES

The next overlap area in figure 1.1 is that of feminist science studies, which amalgamates feminist studies and science and technology studies. The relationship between these two is aptly characterized by the term "unsettled relations" (Thornham 2000).

Donna Haraway, interviewed for this volume (see chapter 3), stresses that feminists like herself and other pioneers of feminist science studies have been "hard to digest" for mainstream chroniclers of the history of science and technology studies. The point she makes is more or less parallel to the one Thornham (2000) has made about the relationships between feminism and cultural studies. There are many overlaps and areas of common ground between feminist science studies and science and technology studies, but that does not mean that the former is just a chapter in the academic history of the latter.

Let us look first at the overlaps. Like feminist studies and cultural studies, both science and technology studies and feminist science studies can be described as having emerged from a founding act of subversion. Somewhat in parallel to the subversion of notions of elite culture, feminist science studies and science and technology studies have critically destabilized the positivist notion of science as a logically self-developing entity, one that was often translated into rational sequences of technological applications that determined the progress of societal development. This destabilization has been called a critique of technological determinism. A joint target of this critique is the natural sciences' hegemonic claim to be able to set the standards for all ways of constructing scientific explanations. Constructivist reinterpretations of technoscience were gradually developed in the wake of discussions about the intertwinement of science and society, discussions provoked by Thomas Kuhn (1962), among others. Technoscience came to be seen as one social and cultural activity among many others. Later, from a deconstructivist perspective, science came to be defined as a material-semiotic practice (Haraway 1991a, 1991b, 1997). Technoscience had to be studied as a part of sociotechnical networks or actor networks, which could be understood as "co-constructions" of human and nonhuman actors (Haraway 1992: 296).[4]

The field of feminist science studies, although it shares these frames of reference with influential trends in science and technology studies, also has its own established genealogies. As described in more detail elsewhere (Lykke and Braidotti 1996), one source of feminist science studies was the political critiques and activism of the women's movement in the 1970s, of which the famous volume *Our Bodies, Ourselves* (Boston Women's Health Book Collective 1971) was an early milestone. That book gave rise to feminist self-help groups in many countries. Another important source was the critique of the ingrained phallogocentrism of the natural, biomedical, and technical sciences, a critique that feminists in those disciplines began to articulate by the late 1970s (Keller 1992), in much the same way that feminists in the humanities and the social sciences had done somewhat earlier. These critiques drew inspiration from and contributed to the development of social constructivist approaches to science. This was the main basis for the gradual intertwinement and overlap between feminist science studies and science and technology studies. A main point of difference between these two fields lies in feminism's introduction of a gender perspective into constructivist technoscience and in feminism's focus on both the embodiment and the social and cultural situatedness of the scientist (Keller 1985; Harding 1991; Haraway 1991a, 1991b).

SCIENCE STUDIES AS CULTURAL STUDIES: INTERSECTIONS BETWEEN CULTURAL STUDIES AND SCIENCE AND TECHNOLOGY STUDIES

The label of the third field of overlap in figure 1.1—"science studies as cultural studies"—is taken from Haraway (1992), which in turn was a contribution to *Cultural Studies* (Grossberg, Nelson, and Treichler 1992), the key collection that brought the label "cultural studies of science and technology" onto the Anglo-American scene. (The genealogies of this emergence are described in more detail in chapter 2.) Since 1992, that label has proliferated to cover many different positions and approaches. This field is somewhat different from the ones just described because is has not been as much a part of community building. An inter- and transdisciplinary community organized around cultural approaches to technoscience can be identified, but under a somewhat different name. The international Society for Literature and Science (SLS), founded in 1985, does represent such a community, to some extent, since "literature," over the years, has come more and more to stand for "culture" in a broader and more cultural studies–like sense. In fact, the SLS has discussed substituting "culture" for "literature," but a recent change of name did point in another direction: instead of substituting "culture" for "literature," the term "arts" was added, a move whereby the close connection to "high" culture was underlined rather than loosened (the organization is now known as the SLSA).

Even though cultural studies of technoscience, as a field, is less institutionalized than the other spaces in figure 1.1, it is still possible to map out some of its significant features, on the basis of the fact that the label "cultural studies of technoscience" has proliferated. I will follow the logic of my tour through figure 1.1 and point out founding acts of deconstruction, relevant to an understanding of the field's key dynamics and its contributions to the emergence of feminist cultural studies of technoscience.

One such act involves a focus on two-way links between technoscience, on the one hand, and literature, the visual arts, fiction, fantasy, and rhetoric, on the other. To lay claim to these links between the formations of technoscience and the cultural imaginary is to undermine the positivist notion of science and technology as radically as does the approach of science and technology studies itself. "The cultural imaginary" is a term used by cultural studies and feminist scholars in referring to intersections between phantasmatic images and discursive forms, in which cultural communities mirror and articulate themselves, and which act as points of reference for collective identity formations. The intersection of the social and the psychological is an important aspect of this concept (Dawson 1994; Bryld and Lykke 2000; Franklin, Lury, and Stacey 2000). To link the cultural imaginary with science is effectively to break down the barriers set up by positivist epistemology between subject and object, and between the emotions and reason.

As a field, cultural studies of technoscience also undermines the classic elitist notion of culture. It is discarded in order for the focus to shift to such signifying practices as popular science, media representations of science and technology, science fiction, the lived experience of technobodies, the technological practices of everyday life, and so on.

FEMINIST CULTURAL STUDIES OF TECHNOSCIENCE

In my tour through the spaces of overlap in figure 1.1, I have now come to the area in the middle, the space where the figure brings all the others together to intersect, amalgamate, and hybridize. This area is labeled "feminist cultural studies of technoscience."

Feminist cultural studies of technoscience, as a field, brings together the founding acts of deconstruction of all the other overlapping areas. Thus it connects the displacement of positivist notions of technoscience, the discarding of beliefs in technological determinism, and the explosion of elitist notions of culture. In so doing, it makes room for an analysis of technoscience as cultural activity, as signifying practice, and as everyday life practices. At the same time, this perspective is linked with a feminist approach, which is not limited to the study of sex/gender issues but instead fits the description of a special kind of hermeneutical tradition that can "open up" an unlimited number of topics. This point is stressed by Franklin, Lury, and Stacey (2000: 6): "Indeed, it is increasingly clear that there are no topics or phenomena to which a feminist analysis is not relevant—at which point it is useful to consider feminist theory as a hermeneutical tradition, or as a set of techniques, rather than as a fixed set of positions or models."

Feminist cultural studies of technoscience also creates a hybrid of different actors' perspectives. Like feminist studies and cultural studies, it situates itself within the perspectives and positions of inappropriate/d others. In the list of actors marked by gender, race, ethnicity, class, and sexual identity—i.e., human actors whose positions have been analyzed in different ways by feminist studies and cultural studies—feminist cultural studies of technoscience may also include the perspectives of nonhuman actors. Their inclusion is in line with studies in science and technology that have given much attention to the significance of the activities of nonhuman actors in sociotechnical networks.

Indeed, some feminist scholars in the field of science and technology studies have radicalized theories of the position of nonhuman actors. Probably the most famous example of this radicalization is Haraway's appropriation of the cyborg (Haraway 1991a, 1991b). As mentioned in the introduction to this volume, the term "cyborg" is an abbreviation of "cybernetic organism." In a more popular vein, we may speak of technobodies: machine-humans, or machine-animals. Haraway, referring to the fusions and to the increasingly intimate cohabitation between bodies and machines that are taking place in the wake of the current digital and biotechnological revolutions, propelled the cyborg into feminist theory as an embodiment of the blurred boundaries between nature and culture, machine and human, matter and mind, sex and gender. The cyborg—partly human/animal and partly machine, partly organic body and partly technological artifact—is a figuration that signals a collapse of the central dichotomies on which positivist science is based. Blurring the boundaries between human and nonhuman, the cyborg figure can be used to mobilize possible connections between the human others of feminist studies and cultural studies, but also the interplay between human and nonhuman actors that actor-network theory may focus on. In Haraway's work, cyborgs—or dogs, in her most recent book (Haraway 2003)—are subjectivized along these lines. Cyborgs are generally treated as cultural and political subjects; nonhuman actors, such as animals, technobodies, or the so-called material world, are ascribed a kind of subjectivity as "witty agents" (Haraway 1991b: 199). Instead of the focusing exclusively on human actors as bearers of alternative political and cultural subject positions, Haraway creates synergy among feminist studies, cultural studies, and actor-network theory by enrolling cyborgs and other nonhuman actors in the ethically and politically affiliated "lumpenproletariat" of "inappropriate/d others."

CONCLUSION: BETWEEN CYBORGS AND BITS OF LIFE

Haraway's feminist appropriation of the cyborg is perhaps the most influential figuration produced by feminist cultural studies of technoscience. The cyborg has been used for very many purposes in feminist theories—so many that it has lost some of its critical power. Haraway has commented on the way in which her cyborg figure caught the imagination of her many readers, and she emphasizes that "as an oppositional figure the cyborg has a rather short half-life, and indeed for the most part, cyborg figurations, both in technical and in popular culture, are not, and have never been, oppositional, or liberatory, or had a critical dimension in the sense that I use critique, i.e., in the sense that things might be otherwise" (cited in Lykke, Markussen, and Olesen 2004: 326). But even though Haraway clearly states that the critical cyborg turned too dizzy with success, she sticks to the idea of the cyborg as a productive methodology or "thinking technology" (ibid.: 335).

The "bits of life" figure, which is the central pivot of this book, can in many ways be seen as a follow-up to or daughter of Haraway's cyborg. Referring both to infotechnological and biotechnological processes of bodily redesign and convergences, the "bits of life" figure also points in the direction of the blurred boundaries between organism and technology, and between matter and discourse, as described in the introduction to this volume.

Nevertheless, with an intensified focus on provisional cuts rather than on bounded entities, the "bits of life" figure may also increase the problems of the cyborg while enhancing its potentials as a feminist thinking technology. As stressed in "A Cyborg Manifesto" (Haraway 1991a), the cyborg is a double-edged figure, apt both for feminist opposition and for teaming up with and sustaining the power of traditional technoscience and its social relations. The same can be said about the "bits of life" figure. Like the cyborg, "bits of life" can be seen as a figure that can both act alternatively and lead in highly problematic directions. At one end of the spectrum, for example, the "bits of life" figure may signal the breaking down of bodies into genetic "components" and conjure up the horrors of a world in which class segregation is constructed around access to the genetic enhancement of bodies, as portrayed in the film *Gattaca* (see chapter 7, this volume). At the other end of the spectrum, however, "bits of life" may be used as a thinking technology that may help reconfigure bodies and worlds as fluid and relational processes, enabling new becomings and new kinds of empowerment of "inappropriate/d others" (Haraway 1992; Minh-ha 1986–87), as theorized by Rosi Braidotti, among others (see chapter 12, this volume). Basically, the field of feminist cultural studies of technoscience is concerned with critiquing the former dimensions and strengthening the latter, affirmative dimensions in an open-ended, transdisciplinary process of disentangling social, cultural, technological, and biological power relations. In this sense, the field can also be characterized as cyborg studies, or as knowledge production on "bits of life."

I conclude this chapter along the lines of the affirmative dimensions, expressing the hope that the "bits of life" figure, as a daughter of the feminist cyborg, may help to strengthen a new dimension in feminist cultural studies of technoscience. As underlined in a recent volume on feminist science studies (Mayberry, Subramaniam, and Weasel 2001), the development of a new feminist relationship to technoscience seems to be under way. The contributors to this volume suggest that the point is no longer to focus exclusively on the critique and deconstruction of prevailing bodies of technoscientific knowledge. This is still highly important, to be sure; but, as they emphasize, today they are also committing themselves more and more to alternative—reconstructive—technoscience projects, that is, to projects that represent a new way of doing science, one informed by feminist critiques and epistemologies. It is to be hoped that the "bits of life" figure, precisely because of its strengthening of the focus on provisional cuts and its relentless openness to new becomings, may be helpful in this process.

NOTES

1. An earlier version of this chapter was published in Braidotti, Nieboer, and Hirs (2002).

2. A Venn diagram is a model in which two or more circles are brought to intersect so as to describe and compare attributes and characteristics of things, people, places, events, ideas, and so on. Venn diagrams are used in many contexts, from preschool education to the highly sophisticated environments of computer science. As a tool for describing and comparing attributes and characteristics, the Venn diagram is a popular device that one can construct interactively at many specially designed Web sites; see, for example, www.venndiagram.com/venn01.html (retrieved March 6, 2007). The virtue of the Venn diagram is that, not unlike the semiotic square, it is so obviously reductive and essentialist that it cannot be taken completely seriously as a vehicle for universal truth values. It is a pedagogical instrument for creating moments of order, and it is nothing more than that. At a deeper level, however, it is, as Haraway writes of the semiotic square, "almost a joke" on "elementary structures of signification" (Haraway 1992: 305). The diagram functions as an engine, pointing out semantic components in such a mechanical and interchangeable way that the user cannot help questioning (jokingly, to be sure) the stability of the whole setup. As with the semiotic square, the enginelike nature of the Venn diagram encourages interactive use, whereby one can keep on substituting the contents of the main circles and getting still new patterns of overlap. An additional virtue of the Venn diagram is that it exposes what Haraway calls a "multi-patterned interaction" representing an "amodern history" (Haraway 1992: 304), a history that evades a modern logic of progress and linear chronology. Neither the semiotic square nor the Venn diagram has a beginning or an end. Moreover, the Venn diagram fits in with the notion of implosion because it is a visual representation of a multipatterned, nonchronological, highly dynamic interaction; contrary to the explosion, with its spectacular beginning, its catharsis, and its end point, the Venn diagram does not impose notions of a linear story line. The diagram also fits Haraway's articulation of the open-endedness of the imploded knot because, as an engine, it noisily asks for substitutions, thus creating new patterns.

3. Because of space limitations, in this chapter I use "feminist studies" as an umbrella term that includes feminist as well as gender and/or women's studies. I will not detail the differences between and among these terms.

4. To avoid collapsing human and nonhuman actors, the tradition of science and technology studies often distinguishes between "actors," who are human, and "actants," which are nonhuman agents. Actants fulfill a function, but without the same kind of intentionality that is characteristic of humans. In this chapter, the term "actor" is used as an umbrella term for human as well as nonhuman actors.

2 ▪ Roots and Routes

The Making of Feminist Cultural Studies of Technoscience

MAUREEN McNEIL

For many of the readers of this volume, the term "cultural studies of technoscience" may be rather familiar. Until fairly recently, however, cultural studies and technoscience studies were not commonly linked. This chapter is concerned with the trails that have constituted the "cultural turn" in science and technology studies.

There have been some conferences and publications that can be cited as markers in the emergence of this field. For example, the proceedings of a 1990 conference at the University of Illinois led to the publication of Grossberg, Nelson, and Treichler (1992), with a section on science, culture, and the ecosystem. A 1994 conference at the Center for Cultural Studies at the City University of New York was much more specifically focused on science and technology. The volume (Aronowitz et al. 1996) that emerged from this conference opens with nothing less than a "manifesto" on the "cultural study of science and technology" (Menser and Aronowitz 1996).

Definitions of cultural studies of science have themselves proliferated and range from vague ones, which could embrace virtually any form of science studies, to sweeping epistemological claims. Rouse (1992: 2) chose to define the term "broadly" so as to "include various investigations of the practices through which scientific knowledge is articulated and maintained in specific cultural contexts, and translated and extended into new contexts." The distinctive element here is Rouse's highlighting of the work—"the practices"—required in the articulation of scientific knowledge, and his emphasis on its being context-dependent.

Menser and Aronowitz (1996: 16), by contrast, are quite dramatic in their declaration that cultural studies is "the name we give to the transformation of social and cultural knowledge in the wake of an epochal shift in the character of life and thought whose origins and contours we only dimly perceive." Aronowitz (1993: ch. 7) is even more sweeping in his claims for the field of cultural studies, heralding it as the "new paradigm" for knowledge, which insists (in contrast to the natural sciences) on the contextuality of all knowledge claims. Nelkin (1996: 34), an established (and now deceased) science studies researcher, observes that some humanities researchers and social scientists have recently taken to "defining their work as 'cultural studies of science' and bringing to bear their skills in interpreting narratives and discourses."

Hence even in a single collection (Aronowitz et al. 1996) there are very different versions of "doing" cultural studies of science, from a rather modest "add literary techniques to social studies of science and stir" approach (Nelkin 1996) to a vision of an epistemological revolution (Menser and Aronowitz 1996). The three recent definitions of cultural studies of science cited here are indicative of the instability of and diversity in this field. I have chosen them because they illustrate three key dimensions of cultural studies of technoscience: epistemology, methodology, and disciplinarity/transdisciplinarity.

It is not my intention to adjudicate these definitions but instead to register the uncertainty about the doing of cultural studies of technoscience, and the profound dimensions of the questioning that this list suggests. Against this background, I am interested in tracking some moments in the emergence of the field and reflecting on its contours, its orientations, and its political significance, particularly for feminism.

The exploration of cultural studies of technoscience presented in this chapter is to some extent in dialogue with the related project undertaken by Lykke (2002; see also chapter 1, this volume). Lykke maps feminist cultural studies of technoscience as an implosion of the overlapping fields of feminist studies, cultural studies, and science and technology studies. The present chapter, by contrast, is a genealogy that traces some patterns of emergence in analytical practices over time. I want to investigate specific developments involving feminist influence in and shaping of the emerging field of cultural studies of technoscience. Moreover, I am offering here only my own "partial perspective" (Haraway 1991b) with a genealogical account focusing primarily on Anglo-American research. My hope is that my account of some bits of the coming into life of feminist technoscience in Anglo-American settings will encourage reflections about other ways in.

DIFFERENT ROOTS AND DIFFERENT ROUTES

As the preceding brief review indicates, cultural studies of technoscience means different things and may involve different activities. It is nevertheless possible to say something about some routes into this transdisciplinary domain. In this chapter, I consider approaches coming from five different Anglo-American disciplinary/ interdisciplinary roots and routes:

1. Cultural anthropology
2. Literary studies of science
3. Studies of visual culture
4. British cultural studies
5. Feminist science fiction studies

The use of disciplinary and interdisciplinary labels as tracking devices seems appropriate and convenient, since it helps denaturalize the mythical assumption that all research in feminist studies, cultural studies, and technoscience studies is inherently interdisciplinary or transdisciplinary. I want to tease out trajectories toward interdisciplinarity and highlight methodological borrowings and adaptations that are sometimes obscured by this kind of naturalization. (Such charting does not register overlaps, but Lykke maps those in chapter 1). There are some important distinctive features to each of these strands, and they merit attention in this chapter, which also registers fruitful exchanges and cross-fertilizations.[1]

Route 1: Cultural Anthropology

Cultural anthropology has its roots in Western imperial explorations during the nineteenth and early twentieth centuries. From the encounters between Western explorers and colonialists of this period emerged practices of systemic observation, documentation, and characterization of "other" cultures. These methodological traditions were channeled into this discipline focused on ethnography (multimethod immersion in cultures for a significant period, involving fieldwork in a "foreign" culture). The object of such fieldwork was the production of insightful analysis of "other" (non-Western) societies, and thus "culture" emerged as the central focus and key term of this discipline.

Cultural anthropology has been closely associated with colonialism and imperialism and their legacies. From the 1970s through the 1990s, however, an explicit, critical confrontation with these origins fostered a radical turn in the discipline, which opened its resources for science studies. With an awakened awareness of the discipline's colonial links, some cultural anthropologists turned their analytical gaze in the direction of Western culture itself. Controversial though this was, the result was a fresh body of investigation and scholarship that attempted to distill and document features of Western societies and cultures.

Once Western societies and cultures had come under anthropological scrutiny and become the subject for fieldwork investigation, it was almost inevitable that the discipline would begin to examine science as a crucial element of Western culture. Nevertheless, it is important not to underestimate either the controversy surrounding this new direction of the discipline or its significance. David Schneider, Marshall Sahlins, and a few other established figures led the discipline in this direction. But feminist scholars were also in the vanguard of this new branch of anthropological study. Among these scholars, some, including Margaret Lock, Marilyn Strathern, and Helen Verran, had begun their careers doing fieldwork in the traditional anthropological manner: observing non-Western cultures. Their earlier fieldwork experiences informed their subsequent investigations of Western science and medicine. Other feminist anthropologists, mainly from the "next generations" in the discipline, did their fieldwork in the West exclusively.

Feminist scholars have adapted two key methodological and theoretical traditions in anthropology—ethnographic research and kinship studies—in order to investigate science and technology in Western culture. The rich tradition of ethnographic study has been brought back "home" by anthropologists and others who have undertaken recent studies of the daily cultures of laboratories, clinics, and other scientific settings. This strand of feminist cultural studies undertook the investigation of rituals and practices instantiated by technoscience.

Nevertheless, it is feminist work on reproduction and kinship that has been most prominent in this branch of cultural studies of technoscience. Engaged with reproductive politics through feminism, and faced with the development of new reproductive technologies in the late twentieth and early twenty-first centuries, some feminist anthropologists turned anthropological kinship studies in a new, critical direction. The research of Jeannette Edwards, Sarah Franklin, Faye Ginsburg, Emily Martin, Rayna Rapp, and others was exemplary in realizing this reorientation. Davis-Floyd and Dumit (1998) edited a collection that demonstrates the rich vein of feminist studies of reproductive science and technology along this anthropological route.

At the heart of much of the feminist work in this stream of anthropologically oriented science and technology studies has been an ambivalence about the "othering" of cultures that is characteristic of anthropological study. As previously indicated, the acknowledgment of anthropology's entanglement with colonialism and imperialism brought suspicion about its role in "othering" non-Western cultures. On the one hand, the turn to Western science and technology constituted a disavowal of this distancing, bringing anthropology home, as it were. On the other hand, one of the strengths of anthropology was that it tended to render strange ("other") precisely those features of Western culture that are often naturalized. Moreover, critical perspectives, including those associated with feminism, generated a distancing within studies of science and technology. Interestingly, such "othering" of science and medicine has been problematized in some recent critical technoscience studies research (Mol 2003).

Feminist anthropological science studies emerged as part of a reflexive turn in the discipline itself. This turn was informed and shaped by encounters with postmodernism and poststructuralism. One manifestation of this turn was greater exploration of writing as a key part of anthropological practice. This was also a feature of intensified analytical and ethical concerns about the role of the investigator. Feminist scholars have been at the forefront of this development, but they have also been involved in gingering up critical perspectives on ethnography as a rich but problematic methodological tool chest that continues to be unpacked in efforts to transform the power relations of the discipline.

Route 2: Literary Studies of Science

As already indicated, feminist scholars and others turned cultural anthropology on its head in undertaking their cultural studies of science and technology. In ethnography and the tradition of kinship studies they found complex and rewarding sets of research methods and frameworks. Feminist literary scholars brought to science studies a rather different but equally rich repertoire of practices culled from literary and linguistic disciplines. This strand of feminist cultural studies of science was first forged through the scrutiny of scientific texts as cultural artifacts that could be subjected to literary and linguistic analysis. Although initially such approaches were cautiously framed, this research firmly identified science as cultural.

The appearance of Gillian Beer's *Darwin's Plots: Evolutionary Narrative in Darwin, George Eliot, and Nineteenth-Century Fiction* (Beer 1983) was a key moment in the emergence of this strand of cultural studies of science. Three years later came a related anthology (Jordanova 1986). Taken together, these two volumes assert the textuality of science, locate scientific writing in a broad panorama of cultural forms, and subject scientific texts to diverse forms of literary and linguistic analysis.

Beer (1983, 2000) unequivocally brings literary studies into the domain of science studies in a book that has been described as "deliberately 'literary' in its approach" and in which she "sets out to read Darwin as a writer who also happened to be a scientist" (Levine 2000: xi). By locating Charles Darwin's 1859 *On the Origin of Species* on a complex cultural tapestry, Beer (1983, 2000) establishes that his evolutionary theory can be fully deciphered and appreciated only through reference to poetry and fiction. Tracing Darwin's "metaphors, myths, and narrative patterns," Beer (2000: 5) shows the cultural resources, particularly those deriving from an English literary heritage, from which Darwin drew in building his arguments.

Beer's work was a turning point in the emergence of cultural studies of science by way of literary studies. For some, this was a "bridge too far" between the "two cultures" (arts and sciences) that C. P. Snow had identified in the late 1950s (Snow 1993). This is obvious from George Levine's foreword to the second edition of the book, in which he praises Beer while warding off the threat that Darwin's theories will be seen as merely cultural; he insists that Beer has refused "historical or social reductionism" and maintained Darwin's "special genius," and that her study of the language of Darwin's texts "undercuts the implication that Darwin was absolutely a man of his time, explicable in terms of the conventions of the middle-class society to which he so nervously and doggedly adhered" (Levine 2000: xi, x).

Despite the many achievements of Beer's book, there are also strictures on her version of cultural studies of science. Her adoption of a rather Kuhnian (Kuhn 1962) framework in conceptualizing scientific theories imposes some temporal contingency on the contention that science is fully cultural. One comment in the opening paragraph of *Darwin's Plots* is indicative in this respect: "When it is first advanced, [scientific] theory is at its most fictive" (Beer 2000: 1). Here, Beer seems to suggest that scientific theory gradually loses its cultural moorings or associations. Moreover, the book vividly conveys Beer's regard for Darwin as a creative writer and her sense of his genius—his singularity. To some degree, the emphasis on genius, singularity, and high culture countervails and contains the promise of providing a full cultural study of Darwin's theory. Hence it is interesting to find Beer explicitly distancing her project from that of the Darwin biographers Desmond and Moore (1991), who present Darwin as embodying and exemplifying nineteenth-century middle-class English masculinity. In fact, whereas Beer traces the links between Darwin's *On the Origin of Species* and the established English literary canon, Desmond (1984) investigates the connections between Darwinian scientific theory and popular culture.

While her original study (Beer 1983) involves a fairly traditional literary and high-culture version of cultural studies, without reference to gender relations, sexuality, or feminist perspectives on science, Beer subsequently linked her work to a new wave of scholarship in feminist science studies. In her preface to the second edition of *Darwin's Plots* (Beer 2000), she praises feminist science studies, which she presents as akin to her own earlier project. Although the second edition of her book is not a revised edition, Beer does wistfully indicate her wish to have been able to draw on that scholarship in her own study. Between the first and second editions of Beer's landmark text, other feminists used and extended literary frameworks in diverse directions. *Body/Politics* (Jacobus, Keller, and Shuttleworth 1990) is perhaps the best-known publication of this sort, although it is much more interdisciplinary than Beer's study. Discourse is its linking concept, the female body in biomedical discourse is its predominant focus, and psychoanalysis figures in a number of the contributions to this explicitly feminist collection.

Route 3: Studies of Visual Culture

The literary orientation of the strand of cultural studies of science forged by Beer (1983, 2000) and Jordanova (1986) is striking. The next major project by Jordanova (1989) sets out in a rather different direction. In that work, Jordanova embarks explicitly on gender and sexuality studies, informed by feminist scholarship, but she also unleashes cultural studies of science from its exclusively literary or linguistic moorings. In addition to exploring literary texts, she analyzes paintings, sculptures, and scientific diagrams to provide an interesting set of studies on gender relations in biomedical science between the eighteenth and twentieth centuries. By using the analytical repertoire of art history to bring science's visual dimensions into cultural studies of science, Jordanova complexifies understandings of the cultural workings of science. She categorically and methodologically extends and enriches cultural studies of science as the study of high culture.

Jordanova's explorations of the visual dimensions and resonances of science influenced a group of feminist scholars working in film studies, who shaped a distinctive trajectory in cultural studies of science. Two special issues of the feminist film studies journal *Camera Obscura*, both titled *Imaging Technologies, Inscribing Science* (Treichler and Cartwright 1992a, 1992b), effectively announced this new subfield and assembled exemplary studies. These issues were edited by Paula Treichler and Lisa Cartwright, but the contributors included Ann Balsamo, Giuliana Bruno, Valerie Hartouni, Constance Penley, Ella Shohat, and Carol Stabile, who all became associated with feminist cultural studies of technoscience.

There were a number of distinctive features of this strand of feminist cultural studies of science. These two special issues of *Camera Obscura* had been inspired by political activism, and the editors regarded their research as a political intervention in its own right. Their introduction linked these two issues of the journal with "a new wave of activism toward institutionalized science and medicine" (Treichler and Cartwright 1992c: 5) that was associated particularly with AIDS activist groups (such as ACT UP) and the National Black Women's Health Project in the United States.

This explicitly political orientation informed and shaped the research presented in these two issues of *Camera Obscura*. The pivot of much of the research was the identification of imaging that was common to science and popular film, such as the demonstration of links between the clinical or scientific gaze and the cinematic gaze. The editors observed, "The technological history and culture of film and television repeatedly overlap and intersect with developments in scientific imaging; science, in turn, uses popular imaging conventions even as it remains emphatically aloof" (Treichler and Cartwright 1992c: 6). Moreover, the contributors to these issues of the journal were also mindful that "many imaging technologies are targeted and have special implications for women" (ibid.: 8), and that investigations of such imaging would necessarily entail gender analysis. These special issues of *Camera Obscura* contained articles that became, in effect, trailers for later books published by Cartwright and Bruno, who have been important figures in bringing film studies into cultural studies of science. In rather different but complementary projects, Cartwright (1996) and Bruno (1993) explore the intersections between the genealogies of physiology and anatomy and the emergence of cinema. In both cases, the author identifies the female body as a key focus in the development of these different and powerful cultural technologies.

Cartwright (1996) undertakes a set of detailed case studies of the use of visioning technologies in the development of early twentieth-century physiology. Bruno (1993) investigates the emergence of popular cinema in early twentieth-century Naples, arguing that its forms of spectatorship are derived directly from the practices of anatomical science. She regards psychoanalysis as a new human science that emerged from the replacement of anatomical performance and the popular focus on the figure of the hysterical female body. Referring to the screening of *La neuropatologia* (a film made by Dr. Camillo Negro in 1908, and shown in Turin that year), she comments: "The audience who customarily attends the performance of the anatomy lesson or watches the theater of hysteria becomes institutionalized as spectatorship. . . . *La neuropatologia* transforms the anatomic table into both filmic screen and psychoanalytic couch and points to the female body as the ground upon which this transformation is enacted" (Bruno 1993: 75). Cartwright, Bruno, and others, noticing the common focus on the body, particularly the female body, that bonds the biological sciences, scientific medicine, and psychoanalysis to cinema, and observing the importance of the use of visual technologies in the development of physiology, were launched (as others also were) on investigations into what they labeled the "common visual language of these fields"(Treichler and Cartwright 1992c: 6). They also noted the increasing use and importance of visualizing technologies in scientific and medical practices. Hence the centrality of media was registered and investigated to enrich understandings of the cultural dimensions of science. Moreover, feminist analyses of the female body as spectacle became central to this strand of cultural studies of technoscience.

Route 4: British Cultural Studies

Cultural studies emerged as a distinctive interdisciplinary field of research and education in Britain in the 1970s and 1980s. The Centre for Contemporary Cultural Studies at the University of Birmingham was a key institutional site for this emerging development, which drew on the disciplines of English literature, sociology, and history in studies of the making of meaning in everyday life in Britain. Linked to the New Left politics of the 1960s and 1970s, British cultural studies was challenged and, to some extent, reconfigured in the wake of feminist and antiracist activism and research. Moreover, it was enriched methodologically and theoretically through encounters with structuralist theory and Marxism, and in later incarnations it adapted or responded to poststructuralist and postmodern theories.

The literary orientation of early work in British cultural studies established a tradition that was surprisingly respectful of the English division between the "two cultures" (Snow 1993) of the arts and humanities and the natural sciences. Moreover, there were other factors that militated against the possibility that science and technology studies might come under the scrutiny of cultural studies researchers in Britain during those years. Marxist scientism cast its shadow strongly in this field. Louis Althusser was one of the main theoretical influences in cultural studies during the 1980s, and his contention that there had been an epistemological break between Karl Marx's prescientific writing and his scientific writing in *Capital* was indicative and influential. More generally, the preoccupation with popular and alternative culture(s) among these cultural studies researchers distracted them from the study of science, which was identified with "high" culture. Indeed, in a recent study of research methodologies and the tradition of ethnographic research in British cultural studies, Gray (2003) has shown an effective aversion to "researching up," which also contributed to neglect of the world of science.

The interest in developing fresh perspectives on popular culture, and on positively investigating active relationships to the mass media, also estranged many cultural studies scholars from the main critical, "continental" theoretical tradition, which did focus on science: the Frankfurt School. In fact, in many respects, British cultural studies became antithetical to the perspectives of the Frankfurt School because that group was identified with negative evaluations of popular culture, associated with the emergence of mass media. This foreclosed the possibility of engagement with its critical perspectives on the legacies of the Enlightenment, scientific rationality, and technological progress.

Raymond Williams was the only "founding father," and one of the few prominent figures in British cultural studies, who investigated scientific and technological dimensions of popular culture. Williams's fascinating genealogy of the "ideas of nature" in British culture (Williams 1980), and his investigation of the shifting boundaries between nature and culture (Williams 1973), were somewhat anomalous but still influential projects in British cultural studies. Williams (1974, 1989) also argued forcefully that researchers in media studies should register and explore the technological dimensions of the mass media.

By the late 1980s, the neglect of science and technology appeared as a glaring lacuna in British cultural studies, one that feminists and others were keen to redress (McNeil and Franklin 1991). In the wake of the feminist "body politics" of the 1970s and 1980s (Thornham 2000), critical investigations of the constitution of "the natural," and of the operation of scientific and medical discourses, appeared crucial to understandings of the everyday life of women (and men) in Western societies. The winds of theoretical change ushered in by poststructuralism and postmodernism seemed to be blowing in a similar direction, given that ideas of progress and the grand narratives of science were brought under scrutiny by these theoretical movements.[2]

As Althusser's influence waned, Foucault became a more influential figure. Foucault's interests in the clinic, in biopower, and in various aspects of bodily discipline and medical and scientific discourses bolstered the theoretical ballast for new kinds of feminist cultural studies of science and technology. Outside the academy, social and political controversy that was focused on technology, science, and medicine was at the center of new social movements, such as the antinuclear and ecology movements, as were campaigns around HIV/AIDS. As the field of British cultural studies was reshaped during the 1980s by feminism and by black British antiracism and sexual politics, science and medicine loomed large in the mapping of forms of sexism and racism (Barker 1981). Moreover, researchers attuned to the changes in daily life in the Western world observed the rapid developments in communication and information technologies as well as in the so-called new reproductive technologies in the last decades of the twentieth century.

Changing political priorities, shifting theoretical orientations, and technological developments drew some scholars of cultural studies toward the study of science and technology in the late 1980s and 1990s. In the United Kingdom and the United States, this shift was manifested in a cluster of case studies of science and technology in popular culture that appeared in the early 1990s. In the late 1980s, one of the study groups at the Centre for Contemporary Cultural Studies mapped the methodological and theoretical terrain for such interdisciplinary work (McNeil and Franklin 1991). This group of feminist researchers also analyzed a contemporary episode in abortion politics in the United Kingdom: the controversy around the Alton Bill, which had been introduced into the British parliament in the autumn of 1987, with the aim of shortening the time during which abortion could be performed legally. The group, having worked "collectively . . . [on] examining the place of science and technology within cultural studies," approached this case study with a "combination of approaches from feminist and cultural studies" (Science and Technology Subgroup 1991: 147,148). In addition, Penley and Ross (1991) published a collection of essays that brought together feminist studies, cultural studies, and technoscience studies. In their introduction to this collection, the editors emphasize that the essays are "almost exclusively focused on what could be called *actually existing technoculture* in Western society" (Penley and Ross 1991: xii). These grounded studies were offered to counter hyperbolic claims about liberation through, or total domination by, new technology. The cultural studies imperative—to study the creation of meaning in everyday life—led many of these scholars to studies of popular and alternative cultures linked to science and technology, including campaigns around abortion rights (Science and Technology Subgroup 1991), AIDS activism (Treichler 1999), and hackers (Ross 1991).

The focus on "ordinary people," on social and political activists, and members of particular subcultures and countercultures was continuous with established modes of doing cultural studies, but it was rather more challenging in science and technology studies. Here, the locus of investigation was shifted away from scientific theories, scientists, technologists, and even scientific texts and scientific or technological artifacts. Such cultural studies of technoscience were far more interested in analyzing the dispersed and diverse creations of meaning around science and technology in popular culture. The heretical implication of this new kind of technoscience studies was that scientists and technologists did not control and could not regulate the meanings of technoscience.

This strand of cultural studies highlights the notion that an understanding of technoscience necessarily involves studying the multiple, diverse, and complex interactions with science and technology that mark daily life in the contemporary Western world. This area of the field is also characterized by specific methodological and theoretical adaptations of British cultural studies. One example is the employment of the concepts of subculture and counterculture. Penley and Ross (1991), for instance, in addition to their joint work, were also undertaking individual studies of specific subcultures and countercultures. Penley (1997) analyzed *Star Trek* and its followers, and Ross (1991) studied environmental groups in the contemporary United States. These methodological and theoretical features are consonant with the political orientation of this version of cultural studies of technoscience, which contends that we are all active in making meaning in contemporary science and technology.

Route 5: Feminist Science Fiction Studies

Second-wave Western feminism generated a distinctive body of imaginative writing that conjured and explored encounters with technoscience. Feminist science fiction, which appeared in the wake of the women's movement of the 1960s and 1970s and of the gender and sexual politics of the last decades of the twentieth century, bent and transformed this established genre of popular literature (Rosinsky 1984; Lefanu 1988; Armitt 1990; Roberts 1993; Rose 1994: chap. 9; Donawerth and Kolmerten 1994; Donawerth 1997).[3] Indeed, since Mary Shelley's *Frankenstein* is taken to be one of the earliest science fiction texts, feminism's "golden age of SF" (Rose 1994: 209), in the last quarter of the twentieth century, could be viewed as a feminist reappropriation of the genre.[4] As a cultural form, science fiction was a distinctive imaginative space in which writers and readers could explore aspirations and fears pertaining to the increasing technologization of Western societies.

Donna Haraway celebrated this explosion of feminist science fiction during the late twentieth century. She registered the significance of the flourishing of this cultural form, welcoming the new feminist science fiction writers of the 1970s and 1980s as "our story-tellers exploring what it means to be embodied in high-tech worlds" (1991b: 173). Beyond this, Haraway incorporated science fiction into her studies, inviting the field of technoscience studies to recognize that science fiction is a crucial cultural site in the creation of meanings around science and technology. Her "Cyborg Manifesto" (Haraway 1991a; first published in 1985) illustrates the promise of feminist science fiction as a resource for critical research in technoscience studies.

Haraway herself demonstrates that complex cultural analysis is required for an understanding of the making of modern science and technology. In this respect, a very indicative achievement is her having combined studies of such diverse sites as museums, advertising, popular film, scientific texts, biographies, and autobiographies in the production of her account of the emergence of the twentieth-century science of primatology (Haraway 1989). Conceptually and methodologically, however, one of her most important and relatively neglected contributions to the field of social studies of technoscience is her having brought it into dialogue with feminist science fiction.

Others have followed Haraway's lead, bringing together feminist science fiction and critical analytical studies of technoscience in exploratory ways. Flanagan and Booth (2002) is one such project. The editors explain that their volume "brings together women's fictional representations of cyberculture with feminist theoretical and critical investigations of gender and technoculture"—in fact, each of the individual contributions is starkly labeled "fiction" or "criticism"—and the editors frame their project cautiously: "Rather than collapsing fiction and criticism or sharply separating practice and theory, this collection provides a variety of viewpoints from which to consider . . . the effects of profound and rapid technological change on culture, particularly in women's lives" (Flanagan and Booth 2002: 1–2). The editors and many of the contributors to the volume have recognized, as Haraway also does, that science fiction is "a vital source of narratives through which we understand and represent our relationships to technology" (ibid.: 2.).[5] This collection has proved to be a rich resource for those working at the intersection of technoscience studies, gender studies, and cultural studies.

CONCLUDING CODA

In this coda, I will offer some tentative reflections on my genealogical mapping—my bits of this life—by considering commonalities and variations in the versions of cultural studies of technoscience that I have traced, and then I will return to the issue of feminist affiliations. There are a number of common elements threading through the diverse versions of doing cultural studies of science that I have sketched in the foregoing account. Most obviously, all of the approaches I have highlighted share perceptions of science as culturally produced. As I have indicated, linking science with other cultural forms (such as popular film), or identifying and analyzing specific cultural forms within science itself (such as the scientific text), has facilitated this recognition. The insistence on the cultural specificity of science has directly challenged assumptions about science as universal and transcendent. Moreover, this insistence has generally been accompanied by an increased awareness of science as the product of Western and modern or postmodern culture, albeit a product that has been transported and translated into other global locations and thereby transformed.

This attention to specificity has been accompanied by a greater sense of the complexity of the making of science among cultural studies researchers. A glance across the range of research and investigation covered in this chapter makes it apparent that doing cultural studies of science can encompass the study of a wide range of actors and sites. For many of the researchers mentioned in this chapter, scientists are by no means the only agents in the making of science.

In many respects, the emergence of cultural studies of science has been a methodological explosion within science studies. As I have indicated, researchers have borrowed extensively from the methodological and theoretical tools of anthropology, literary studies, art history, film studies, British cultural studies, and science fiction. Inevitably, these borrowings have been selective. The melding of kinship studies in the investigation of new reproductive technologies is perhaps the most obvious example of this. Moreover, such studies have inevitably adapted and transformed methods and theories, including those of anthropological kinship studies.

In virtually all of the work considered in this chapter, researchers have drawn attention to the mediated nature of scientific knowledge and reminded us that no cultural form—including those forms in science—is transparent. As a result, the field of science studies in general has become much more inclined recently to pay attention to writing, images, and other visual renderings. Unpacking the detailed forms of mediation has been a key methodological thread in the cultural turn of science studies.

While the features just cited are more or less shared elements of the various feminist cultural studies routes traced here, these trajectories also diverge in notable ways. As I have delineated, some of these researchers retain the traditional focus on scientists or scientific communities, taking their texts and artifacts as the starting point for science studies. Others are more interested in the diffused and dispersed generation of meanings around science and technology, investigating everyday, "lay" encounters with science and technology, whether through medical treatment or ecopolitics or *Star Trek* fan groups.

In general, cultural studies has developed an awareness of the historical significance of the high culture/low culture divide and has sought to undermine that distinction. This preoccupation poses interesting dilemmas for those exploring the world of science, with its established high-culture affiliations. Some researchers have respected and, indeed, reinforced that characterization of science. As noted, Beer (1983, 2000) exemplifies this orientation. In contrast, much in the field of feminist cultural studies of science has challenged this distinction, highlighting the multiple agents and agencies involved in the making of science in everyday life.

My broad-brush review of ways into cultural studies of technoscience has also taken account of the field's methodological and theoretical diversity. Textual analysis, visual analysis, and ethnographic research are among the approaches that have been employed. The texts that have emerged from cultural studies of technoscience have themselves taken multiple forms, including thick description, contextual accounts, ethnographic studies, and theoretical studies. Indeed, the theoretical borrowings undertaken by those working in this field have been broad and eclectic and by no means restricted to feminist frameworks.

In addition, these traditions may be distinguished in terms of whether or not they are concerned with investigating the power dimensions of technoscience. Those who have incorporated power as a dimension of their investigations have used different frameworks—Marxist, feminist, Foucauldian, poststructuralist, postcolonial, or various combinations of these. Haraway's work is a case in point, since it is critically orientated toward the power relations of technoscience while her studies employ a range of theoretical resources drawing on all of these frameworks. Some of the work cited here is reflective, methodologically and analytically, about the position of the technsocience studies researcher. In other cases, researchers have adopted rather than repudiated the scientific norm of ostensible value neutrality.

Science traditionally has been seen as a framework for knowledge production, and this feature predominates in the history of social studies of science. In many respects, cultural studies of science have pushed beyond the cognitive dimensions of science. The interest in science fiction and the various detailed research projects about science and technology in everyday life, tracked above, are indicative of this push. Science fiction explores dreams and nightmares around science and technology in the modern world. For many researchers, doing cultural studies of technoscience involves acknowledging that cognitive encounters are only one way in which modern technoscience is lived.

In this chapter, I have contended and demonstrated that the emergence of cultural studies of technoscience has been closely entwined with the history of late-twentieth-century feminism. In virtually every strand of this field of critical scholarly research, feminists have been in the vanguard, and feminist activism has inspired fresh insights and new orientations in approaches to modern technoscience. In this sense, the field of cultural studies of technoscience has not been a magical discovery. Nevertheless, it has been quietly transformative of contemporary understandings of technoscience and insistently optimistic about the possibilities of transforming both science and society. I hope to have shown some of the ways in which feminism has brought new bits of life to technoscience studies.

NOTES

1. For example, some of the work of Constance Penley, Andrew Ross, and Sarah Franklin blends elements of the traditions of cultural anthropology and British feminist cultural studies.

2. Nevertheless, as Penley and Ross (1991) have argued, there are strands of postmodernist theory that are uncritically celebratory in their attitudes toward technology.

3. The list of feminist-influenced science fiction writers of this period is extensive and would include (to name but a few key figures): Margaret Atwood, Marion Zimmer Bradley, Olivia Butler, Suzy McKee Charnas, Zoe Fairbairns, Sally Miller Gearhart, Ursula Le Guin, Anne McCaffrey, Vonda McIntyre, Naomi Mitchison, Marge Piercy, Joanna Russ, Pamela Sargent, and James Tiptree Jr.

4. Rather than Mary Shelley, Rose (1994: 209) nominates as the original "foremother" of science fiction Margaret Cavendish, the seventeenth-century English philosopher and duchess and author of the 1688 utopia *The Description of the New World Called the Blazing World.*

5. Flanagan and Booth (2002) are particularly concerned in this volume with cyberfiction. The introduction, and the chapters by Booth (2002) and Hollinger (2002), provide valuable reviews of cyberfiction as well as fresh perspectives on it and on feminist and queer science fiction more generally.

3 ▪ "There Are Always More Things Going On Than You Thought!"

Methodologies as Thinking Technologies: Interview with Donna Haraway

NINA LYKKE, RANDI MARKUSSEN, and FINN OLESEN

Donna Haraway is a professor in the History of Consciousness Program at the University of California, Santa Cruz. Her theories have been central to the unfolding of cyborg feminism and to feminist cultural studies of technoscience. Major contributions to the field are her books *Primate Visions: Gender, Race, and Nature in the World of Modern Science* (Haraway 1989) and *Simians, Cyborgs and Women: The Reinvention of Nature* (Haraway 1991b) as well as *Modest_Witness* Feminism and Technoscience (Haraway 1997).

Our interview with Donna Haraway was conducted in October 1999, when Haraway visited Denmark as keynote speaker for the conference "Cyborg Identities." It was first published in the Danish feminist journal *Kvinder* (Lykke, Markussen, and Olesen 2000), in a special issue on methodologies. Together with an additional section, it was also published in *The Haraway Reader* (Haraway 2004). With its special focus on methodologies, the following portion of the interview offers an important introduction to frameworks and approaches that have been a major inspiration for the development of feminist cultural studies of technoscience in general, and to this volume.

Interviewer: I would like to start . . . with a question about your writing style. When I teach feminist theory, I often advise the students not only to focus on the line of argument of your texts but also to read them in a literary way, that is, to give attention to the metaphors, images, and narrative strategies and to study how you make the literary moves explicit. I think that you, in a very inspiring way, practice your tenet about "scientific practice" as a "story-telling practice."[1] Your deconstructions of the barriers between theory and literature make your texts extremely rich; theoretical content, methodology, style, and epistemology go hand in hand. How did you come to this kind of writing?

Donna Haraway: Well, there are lots of ways of talking about this. First of all, it is not altogether intentional. Writing does things. Writing is a very particular and surprising process. When I am writing, I often try to learn something, and I may be using things that I only partly understand, because I may have only recently learned about them from a colleague, a student, a friend. This is not altogether a scholarly proper thing to do. But I do that from time to time, and it affects style. It is like a child in school, learning to use a new word in a sentence.

Interviewer: Would you compare this to a literary intuitive way of writing?

DH: Yes. My texts are full of arguments, it must be said [*laughs*]. But my style of writing is also intuitive. It absolutely is. And I like that. I like words. They are work, but they are also pleasurable.

Interviewer: This means that it is possible to keep going back to your texts and still find new inspiring layers of meaning, as in literary texts.

DH: Yes, in a sense, I do think that they are literary texts.

Interviewer: Your efforts to transgress the barriers between theory and literature make me think of other scholars within the feminist tradition—for example, Luce Irigaray—and the ways in which she deliberately links writing strategies and epistemology. Could you tell us how you look upon these links, as far as your own work is concerned?

DH: Well, my style is not only intuitive but also the result of deliberate choice, of course. Sometimes people ask me, "Why aren't you clear?" And I always feel puzzled or hurt when that happens, thinking, God, I do the best I can! It's not like I'm being deliberately unclear! I'm really trying to be clear! But, you know, there is the tyranny of clarity and all these analyses of why clarity is politically correct. However, I like layered meanings, and I like to write a sentence in such a way that, by the time you get to the end of it, it has at some level questioned itself. There are ways of blocking the closure of a sentence, or of a whole piece, so that it becomes hard to fix its meanings. I like that, and I am committed politically and epistemologically to stylistic work that makes it relatively harder to fix the bottom line. When you ask how I came to this, I think that it is actually something I inherited out of my theological formation. In an academic sense, I am trained in biology, in literature, and in philosophy. Those are my academic backgrounds—together with history, but that came later. But I am also deeply formed by theology, and particularly by Roman Catholic theology and practice. I learned it. I studied it. It is deep in my bones. I started reading Saint Thomas when I was about twelve years old, because of the advice of a confessor. It was a way of dealing with doubts about faith. This confessor was a very young priest, a Jesuit, who was ordained with my uncle. He advised me to read Saint Thomas. It was very strange reading for a twelve-year-old girl [*laughs*]. It was very confusing. I did not understand a word [*laughs again*].

Interviewer: But it was your first meeting with layered meanings?

DH: In a way, yes. I had this whole relationship with priests, who themselves were struggling with things. It is a very personal history. Anyway, there was a particular theological frame, which was very powerful for me. Actually, it was not Saint Thomas, but more the whole framework and, in particular, the idea that as soon as you name something and believe in a name, there is an act of idolatry involved—the idea that the names of God are always, finally, deeply suspect, the idea that spirituality has a much more negative quality to it, the idea that if you seriously are trying to deal with something that is infinite, you should not attach a noun to it, because then you have fixed and set limits to that which is limitless, and the whole point of God is about a kind of eternal totality that is not the totality of a system. It is not a systemic totality. It is a different kind of totality. It is an unnameableness. It is the theological tradition that focuses on unnameableness.

Interviewer: Like the Sunni tradition in Islam?

DH: Yes, but there are many traditions that have a commitment to this kind of negativity. There is a strong current within Catholicism that has a commitment to this kind of negativity, and within Quaker practice as well. That is the theological formation that I think is strongest for me—also as regards its relationship to the proofs of the existence of God. These are not about design, and not about causality, but about the reality of infinity, about the truth of limitlessness, which, as I see it, is existent. To me, this is an existentialist idea, and not a design idea. Any proof of the existence of God is almost a kind of joke from that point of view. Well, you know, I am of course a committed atheist and anti-Catholic, anyway at some level. You cannot live in the Christian United States, the right-wing United States, and not be anti-Christian. But that theological tradition is a very deep inheritance for me, and I think it affects my style very deeply.

Interviewer: How do you translate your epistemological and political commitment to the deconstruction of fixed categories into methodologies? How do we avoid fixations? Whenever we are doing research, we use certain sets of skills, which imply certain kinds of names, classifications, categorizations, standards, et cetera. So isn't there a latent or even very active danger, or risk, or possibility that we will always reduce whatever we are doing, even when we have the most ambitious intention about avoiding closure of the discursive spaces in which we theorize, analyze, et cetera?

DH: Well, obviously, there is no final answer to that, because it is a permanent paradox, or dilemma. But there are some things that we can say about that dilemma. First of all, categories are not frozen. We are more inventive than that. The world is more lively than that, including us, and there are always more things going on than you thought; maybe less than there should be, but more than you thought! Secondly, you can use categories to trouble other categories. Marilyn Strathern formulated this very wonderful aphorism: "It matters which categories you use to think other categories with" [*laughs*]. You can turn the volume up on some categories and down on others. There are foregrounding and backgrounding operations. You can make categories interrupt each other. All these operations are based on skills, on technologies, on material technologies. They are not merely ideas but thinking technologies that have materiality and effectivity. These are ways of stabilizing meanings in some forms rather than others, and stabilizing meanings is a very material practice. Thirdly, I find it important to make it impossible to use philosophical categories transparently. There are many philosophers who use cognitive technologies to increase the transparency of their craft. But I want to use the technologies to increase the opacity, to thicken, to make it impossible to think of thinking technologies transparently. Rather, I will foreground the work practice that thinking is. I will stress that category making is a labor process with its own materiality, which is a different kind of materiality than making a sailboat, or raising a dog, or organizing a feminist demonstration. Thinking is involved in all these material practices, but category formation, category manipulation, is a different skill. I do not want to throw away the category-formation skills I have inherited, but I want to see how we can all do a little retooling. This is a kind of modest project, an act of modest witnessing.

Interviewer: To me, your article "The Promises of Monsters,"[2] and the way you use the semiotic square here, are very good examples. Contrary to making your thinking technology—in this case, the semiotic square—transparent and nonconspicuous, you make it visible as a tool, as an analytical technology, as "an artificial device that generates meanings very noisily."[3] I think that this is a very good way of showing how thinking technologies, categories, models, research designs, et cetera, create the object of study.

DH: Yes, I agree with you. I like that reading of "The Promises of Monsters." Here we have this very clutchy structuralist object, and you march through the square [*laughs and beats a march rhythm with her fingers on the table*]. It is a kind of serious joke. I think you can actually do interesting work with these tools, but I want to hear them making noise, I want to feel the friction, I do not want to increase the transparency. Obviously, you have to hold the transparency at a certain level, or you cannot get anywhere. But those are tactical decisions about tools. That is a technoscientific way of thinking.

Interviewer: Your epistemological focus on nonclosure and deconstruction of fixed categories has led you beyond the impasse of standpoint feminism, but it has also been important for you to avoid the traps of relativism and nihilism in which the rejection of stable political grounds has left some postmodern and poststructuralist thinkers. As part of your commitment to situated knowledges, you emphasize the necessity of political accountability. I think that here you point out a very important third path, so to speak, a way of navigating in between pure standpoint feminism and pure postmodern relativism. Do you yourself see your position in this way?

DH: Yes, that is the way I like to understand it, too. I am not looking for the stable ground of standpoint feminism, but nor is my position relativist, nihilist. It is not skeptical. It is not cynical. I believe in limitlessness and nonstable grounds, but at the same time I feel very strongly affiliated with standpoint theorists.

Interviewer: In which ways?

DH: I feel that important work gets done with this very contaminated tool. There are obvious troubles with adopting the metaphors of perspectivalness, of locations and standpoints, of embodiment and privileged perspective, et cetera. I think the contaminations of these metaphors are obvious. But that should not stop us from understanding the crucial work that feminist standpoint theorists did, inheriting [Georg] Lukács, on the one hand, and certain kinds of feminist work that we had been in after all, on the other. People like Nancy Hartsock and others understood standpoint feminism as an achievement. This was an epistemological achievement that came out of a political practice that produced the possibility of understanding the reality beneath the appearances in a specifically Marxist sense—that is, the possibility of understanding the system of domination that supports the appearance of equality; the appearance of normality, and comfort, and equality, and the market, and all of its sequel; the appearance that men and women can simply have a few equal rights and all will be well; the appearance that race can simply be erased by a little bit of antidiscrimination. Standpoint theory produced the understanding of the deep materiality of oppression beneath all these appearances. The method of understanding was the metaphor of surface and depth, which is not the same as making the mistake of misplaced concreteness and [mistaking] analytic technology for the world. The world is not surface and depth. Our analytic technology is about surface and depths. But you do not mistake the analytic technology for the world, because then you would have committed an act of idolatry and fetishism (in the bad sense of the word "fetish"), an act of reification, and this is not what the standpoint theorists did. Such a reading of the standpoint theorists acknowledges their work as an epistemological achievement within a particular intellectual tradition. I find that important, and I also think that this is the way in which the standpoint theorists read themselves. However, the standpoint theorists do not analyze the literary moves of their own texts, because they do not see them as literary moves. But I see them as literary moves, not in a reductive sense. It is not in order to dismiss the texts, but in order to remind myself that this is a set of rhetorical possibilities. This kind of "literary" reading makes the standpoint theorists very suspicious. Actually, I cannot believe the number of people who, in the face of the word "narrative," think that all of a sudden you are *merely* in the realm of culture and entertainment—that all of a sudden you are not talking about what is serious. This is a terrible prejudice, which the standpoint theorists share with most political scientists and a vast majority of philosophers, too.

Interviewer: In a video, *Donna Haraway Reads 'The National Geographic' on Primates*,[4] you visualize your analytical method pedagogically by untangling a ball of yarn. You are pulling out the threads, metaphorically demonstrating a deconstructive move, I guess—a critical going back to where things are coming from. How would you compare this to the Bruno Latour–inspired follow-the-actors approach that I think you are very committed to as well?

DH: Well, I see the pulling out the threads on the video and the follow-the-actors approach as closely related. In my recent book *Modest Witness*,[5] I have this family of entities, these imploded objects—chip, gene, cyborg, fetus, brain, bomb, ecosystem, race. I think of these as balls of yarn, as gravity wells, as points of intense implosion or as knots. They lead out into worlds, you can explode them, you can untangle them, you can somehow loosen them up. They are densities that can be loosened, that can be pulled out, that can be exploded, and they lead to whole worlds, to universes without stopping points, without ends. Out of the chip you can in fact untangle the entire planet on which the subjects and objects are sedimented. Similarly, you do not have to stay below the diaphragm of the woman's body when dealing with the fetus. It leads you into the midst of corporate investment strategies, into the midst of migration patterns in northeastern Brazil, into the midst of little girls doing cesarian sections on their dolls, into the midst of compulsory reproductivity and the question "What is it that makes everybody want a child these days? Who is this 'everybody'?"

Interviewer: How would you describe the relationship between the research subject and the figures that perform in the analysis—for example, your relationship to the figures or imploded knots [called] chip, gene, cyborg, fetus, brain, bomb, ecosystem, race?

DH: Figures are never innocent. The relationship of a subject to a figure is best described as a cathexis of some kind. There is a deep connection between the subject and the figure. It is not just about picking an entity in the world, some kind of interesting academic object. There is a cathexis that needs to be understood here. The analyst is always already bound in a cathectic relationship to the object of analysis, and she or he needs to excavate the implication of this cathexis of her/his being in the world in this way rather than some other. Articulating the analytical object—figuring, for example, this family or kinship of entities [called] chip, gene, fetus, bomb, et cetera (it is an indefinite list)—is about location and historical specificity, and it is about a kind of assemblage, a kind of connectedness of the figure and the subject.

Interviewer: I would like to know about your relationship to science and technology studies, the STS tradition. There are, for example, some obvious parallels between your work and the work of Latour, and he is, in a sense, leaving science studies now. What about you? How do you look upon science studies today? And which role does feminism play here?

DH: Well, science studies is a kind of indefinite signifier, and that is what has made it a good place to locate oneself. It is professionalized in various ways, and that is useful. I will sometimes use science studies as a signifier for myself, and at other times I will not use it. It is a professional and strategic location, but it is not a life-long identity. Even though in some other ways it is, because there are institutional realities connected to it. But people like Susan Leigh Star, and Bruno Latour, and Andy Pickering, and I, and many others—we read each other. So we end up being both deliberately and unconsciously in conversation. But this conversation and reading of each other's texts do not refer to a kind of shared origin story or genealogy. I have a very different genealogy in science studies than, say, Andy Pickering or Bruno Latour do[es]. People like Susan Leigh Star and I share more of a genealogy in science studies, that roots it, for example, in the women's health movement and in technoscientific issues related to women's labor in the office or to Lucy Suchman's work. You know, we share a genealogy of science studies, which, among other things, situates it in relation to the history of the women's movement at least as much as it connects it to a history of a strong program, to a history of actor-network theory, or to a history of a rejection of actor-network theory. You know, all of those end up becoming interesting little events in the neighborhood, but not the main line of action. So in that sense I have a kind of annoyed relationship with some of the canonized versions of the history of science studies, which go like this: "Well, there was this in Edinburgh, there was that in Paris, and whatever." You know, in that narrative of science studies, people like me and my buddies are always hard to incorporate. Even by people of great goodwill, such as Andy Pickering, whom I both admire and read with great pleasure, and like as a human being. None the less, read his preface to *Science as Practice and Culture*[6] and watch the absolute indigestibility of Sharon Traweek and me. We are as the angels with the twelve trumpets. Literally. Every other figure in that introduction got a paragraph or so of analysis, in terms of what was contributed and what he liked or objected to. But we were like blasts from Saint John's Apocalypse [*laughs*]. Literally! That is the figure he used. Because we are not part of that other story in that way of telling it, and they do not know our story. They do not know it as an academic story, and they do not know it as a political story. It is a different history. So after I was already doing what I now call feminist technoscience studies, I read people like, for example, Bruno Latour. So Latour and other authors, [who] figure prominently in the canonized version of the history of STS, were not the origin in my story; they came after other events. And they do not get this—that there is a whole other serious genealogy of technoscience studies! So I remain irritated [*laughs*]. Because we do know *their* genealogies, very well. And they do *not* know ours, even though they exist in writing—our versions of the story are certainly not inaccessible! On the other hand, this does not mean that I would call myself an outsider. That would be silly of me. But I think it remains true in most academic locations, including science studies, that most feminists are both insiders and outsiders in the sense that Patricia Hill Collins theorized this insider/outsider location. Sometimes we are forced into this location, and sometimes we choose to inhabit it.

Interviewer: And I suppose the reason is the issue of feminism.

DH: Yes, we are a little hard to digest. And I think that is a good thing. On the one hand, we are so normalized, and disciplinized, and comfortable, you know, and to call ourselves outsiders is a kind of lie. But, you know, from another point of view, we are still outsiders.

Interviewer: I think that the term you borrow from Trinh Minh-ha in "The Promises of Monsters"[7] describes this position very well. It is the position of the inappropriate/d other.

DH: Yes, you are necessarily inappropriate/d. You know, I am surprised that so few people have used Trinh Minh-ha's term. I agree with you that it is a really good figuration.

Interviewer: When you emphasize that there are other stories about science studies than the canonized ones, I am reminded of the copy of the film poster from *The Matrix* which Don Ihde presented at the conference yesterday as a kind of serious joke, suggesting that the three male figures on the original poster could represent Bruno Latour, Andrew Pickering, and himself, while the only female figure could refer to you. This was Don Ihde's way of jokingly creating a metaphor for the matrix of science studies. But when I saw Don Ihde's matrix, a different matrix of science studies immediately came to my mind. Here, the three male figures were replaced by three female figures—you, Evelyn Fox Keller, and Sandra Harding—while the female figure was replaced by Bruno Latour. This does not mean that I do not recognize the importance of Bruno Latour in the matrix of science studies, but I would simply consider the three other contributors more important in my feminist version of the story of science studies.

DH: Yes, I agree. There are a lot of missing matrices or matricians! Moreover, I can add to the story that many of us have fought with Bruno Latour about feminism, and he has finally been willing to take on something. But it is never symmetrical. He is a friend and a person for whom I have enormous respect. But the asymmetry is a historical, structural problem. It is almost impossible for folks in those locations to get it, and feminist technoscience work always feels like trouble—like "Now you are getting political again."

NOTES

1. Haraway, *Primate Visions* (1989), 4.
2. Haraway, "The Promises of Monsters," in *Cultural Studies*, ed. Grossberg, Nelson, and Treichler (1992).
3. Ibid., 304.
4. Paper Tiger Television (1987).
5. Haraway, *Modest_Witness* (1997).
6. Pickering, *Science as Practice and Culture* (1992).
7. Haraway, "The Promises of Monsters" (1992).

PART 2

■ *Reconfigured Bodies*

4 ▪ Fluid Ecologies

Changing Hormonal Systems of Embodied Difference

CELIA ROBERTS

In her introduction to *Modest_Witness@Second_Millennium*, Donna Haraway suggests that the late twentieth century could be described as an era of "technobiopower" rather than one of biopower (Haraway 1997: 12). In place of Foucault's (1978) nineteenth-century landscape of the administration, therapeutics, and surveillance of bodies as living organisms, Haraway describes a contemporary world constituted by an "implosion of the technical, organic, political, economic, oneiric and textual" that produces cyborg bodies. Such cyborg bodies, she suggests, "inhabit less the domains of 'life,' with its developmental and organic temporalities, than of 'life itself,' with its temporalities embedded in communications enhancement and system redesign" (Haraway 1997: 12).[1] In these contemporary domains, life is "enterprised up" (Haraway 1997: 12), disaggregated into "bits" that are consequently, as Franklin (2003) argues, available for (bio)capital formation. In this world, Haraway suggests, "biology"—referring both to a set of techoscientific discourses and to historically specific forms of lived embodiments—becomes an accumulation strategy: "For us, that is, those interpellated into this materialized story, the biological world is an accumulation strategy in the fruitful collapse of metaphor and materiality that animates technoscience" (Haraway 1997: 97).

For Haraway, Franklin, and many others, this change in the configurations of life and biology has distinct and significant implications for how we live today. Rabinow (1992), Novas and Rose (2000), Rapp (1999), and others use Rabinow's term "biosociality" to describe the ways in which the lives of individuals and groups are mutated through their engagement with contemporary technobiopower. In different ways, all these theorists argue that new biological discourses produce novel identities and forms of social engagement. Inspired by this work and by feminist corporeal theory, I am interested in the implications of contemporary transformations in understandings of biological bodies for theorizations of sexual difference. In what ways do biological bodies continue to "matter," to use Butler's (1993) term, in the production of sexual difference today?

To explore this question, this chapter analyzes a contemporary change in one of Western technoscience's central conceptualizations of the sexed body: that of endocrinology. Critically reading a range of technoscientific, biomedical, and ecological texts, I suggest that the hormonally sexed body is currently undergoing a significant reconfiguration. Throughout the twentieth century, the hormonally sexed body was configured as a bounded, homeostatic system into which pharmacological hormonal interventions could be made straightforwardly. At the turn of the twenty-first century, I argue, this configuration is breaking down. This breakdown stems from two key factors. First, mounting evidence describes the widespread and systemic negative impacts of hormones used as medical treatments (for example, in hormone replacement therapy) on human bodies. The effects of the hormonal antimiscarriage medication diethylstilbestrol (DES) on the descendants of women who took it provide a stark example of the transmission of hormonal effects across generations. Animal evidence is now suggesting that the granddaughters of these women are, like their mothers, at increased risk of reproductive-tract cancers (Newbold, Hanson, Jefferson, Bullock, Haseman, and McLachlan 1998). Second, there is increasing concern about the effects of hormones and chemicals that act like hormones (endocrine-disrupting chemicals) in the environment. These chemicals enter the environment in many ways: they are excreted by humans into water systems; they leach from paints, plastics, detergents, cosmetics, and fertilizers, among other things, into soil, water, and skin; and they enter the food chain when animals that have been treated with hormones or affected by endocrine-disrupting chemicals are consumed by other animals or by humans. The effects of such chemicals are deeply contested but are increasingly described as highly problematic. Through the effects of these two factors, I suggest, hormonally sexed bodies are being reconfigured as unbounded, as lacking coherence, and as under globalized threat from the environment. The contemporary hormonally sexed body, in other words, is disintegrating into "bits."

As an instantiation of technobiopower, this reconfiguration has relevance to thinking not only about hormones and hormonally sexed bodies but also about other aspects of contemporary biological bodies—for example, genes. It also has significant implications for scientific and popular understandings of biological sexual difference and, consequently, for the everyday lives of embodied men and women. In this chapter, I argue that the disintegration of the twentieth-century hormonal body creates significant tensions between, on the one hand, modern models of medical control and intervention and, on the other, a more contemporary version of the body as situated within highly complex ecological systems, the control of which lies far beyond the reach of medicine and must involve governments and corporations as well as individuals in any attempted change. In the case of hormonally sexed bodies, this tension is evident in two ways. First, nongovernmental organizations (NGOs, such as Greenpeace, Friends of the Earth, and the World Wildlife Fund for Nature) have in recent years lobbied governments (at all levels, including supranational bodies like the European Union) and corporations in serious attempts to get them to take responsibility for chemicals in the environment in the face of significant technoscientific uncertainty about their effects. These efforts, as discussed below, are often frustrated. Second (and in the face of significant failures on the part of governments and corporations to respond in ways that will protect citizens from the effects of endocrine disruptors), these tensions feed into increasing demands, voiced most strongly by environmental groups but also repeated in the mass media, for women to become responsible for monitoring their own and other bodies' exposure to hormones and endocrine-disrupting chemicals. Such demands, made in the face of great uncertainty both about the effects of endocrine disruptors and about the possibility of regulation, indicate that the contemporary hormonally sexed body has moved out of the realm of medicine and become an object for gendered practices of care.

TWENTIETH-CENTURY HORMONAL BODIES: HORMONES AS SYSTEMS

Before the nineteenth century, as many historians of the body have argued, anatomists understood the body as a machine that operated according to the laws of physics (Canguilhem 1994). As the new "life sciences" of biology, embryology, and physiology developed in the early nineteenth century (Duden 1993: 103), this view was challenged, and the body instead came to be understood as an elastic, organic machine (Canguilhem 1994: 86).[2] The "birth" of the hormonal body was iconic in this regard. Indeed, as Georges Canguilhem describes it, the French physiologist Claude Bernard's "discovery" of the glucose-producing function of the liver in 1855 was central to the development of the (now ubiquitous) concept of bodily homeostasis. The idea that the body produces chemicals—which Bernard named "internal secretions"—necessary to its own survival was completely new at that time. Indeed, Canguilhem argues that only a few years earlier, the notion of internal secretions "would have been taken as a contradiction in terms, an impossibility as unthinkable as a square circle" (Canguilhem 1994: 265–66). Bernard suggested that the body, as an organic mechanism, was made up of disparate parts that produced chemicals, which kept its internal environment in a consistent state (homeostasis). The body, like the state, contained individual elements that worked alone for the greater benefit of the whole. Moreover—a point important for my argument—the homeostatic body was to a large extent independent of its external environment.

Not until the late 1800s and the early 1900s, as I have described elsewhere (Roberts 2002a), was Bernard's concept of homeostasis applied to what we now know as the sex hormones. The word "hormone" was coined in 1905 to replace the term "internal secretion," which had been used by Bernard and others. The word, coined by the British physiologist Ernest Starling, comes from Greek word *hormōn*, the present participle of *horman*, meaning "to excite or arouse" or "put into quick action."[3] Through this etymology, hormones were figured as messengers: Starling defined hormones as "chemical messengers . . . carried from the organ where they are produced to the organ which they affect, by means of the blood stream" (cited in Oudshoorn 1994: 16).

In the first decade of the twentieth century, European gynecologists applied this idea of hormones as messengers to sexuality and sexual difference (Borell 1985: 12). Somewhat reluctantly, British physiologists accepted that sex could be understood in such terms and that, as a key figure, Edward Schaefer, put it in 1907, it was "highly probable that it is to internal secretions containing special hormones that the essential organs of reproduction—the testicles and the ovaries—owe the influence they exert on the development of secondary sexual characters, and . . . upon the maintenance in a well-developed condition of important internal organs of generation" (cited in Borell 1985: 13). In the early twentieth-century social and conceptual networks of experimental physiology, gynecology, and pharmaceutical biochemistry, then, sex hormones came to be understood as messengers operating within internal chemical feedback systems to maintain bodily sexual homeostasis (Oudshoorn 1994).

In the first half of the twentieth century, as the historian of endocrinology Victor Medvei has demonstrated, this idea of hormones as chemical messengers became linked with explanations of the body's functioning that were current in several other disciplines, including the new field of genetics. Endocrinology, armed with this fertile concept, expanded rapidly, its focus moving away from "the mechanism of the regulation of metabolism, to that of cellular communication by means of chemical messengers" and expanding "far beyond that" in its collaborations with genetics (Medvei 1982: 339). Medvei thus shows that "the clinicians and physiologists were soon joined by the embryologists, biochemists, and even physicists . . . , philosophers and sociologists (on problems of fertility and contraception), not to forget the psychiatrists and neurologists. Even the historians attempted to explain the events of history due to endocrine causes" (ibid.).

In this respect, the twentieth-century hormonal body provides us with a clear (if underdescribed) example of the informational body. As Haraway, Stacey, Martin, and others have argued, the idea of the body as an information system gained widespread acceptance from midcentury on. In this newer model, Stacey (2000: 129) writes, the body increasingly became systematized and understood as a network of communication and was "no longer viewed as a mechanism made up of relatively autonomous, yet interdependent parts." For Haraway, this midcentury transformation began the move from an era of biopower into that of techno-biopower. In this move, nineteenth-century understandings of the biological body as "a labouring system, structured by a hierarchical division of labour, and an energetic system fueled by sugars and obeying the laws of thermodynamics" (that is, the laws of homeostasis) were replaced by understandings of the living world as "a command, control, communication, intelligence system . . . in an environment that demands flexible strategies of accumulation" (Haraway 1997: 97). Also at that time, hormonally sexed bodies began to be detached from understandings of sexuality as stemming from fixed organic origins, and to be reconfigured as feedback loops of biochemical informatics.

By the second half of the twentieth century, then, the hormonally sexed body was understood as an internally homeostatic system of biochemical feedback. Sex hormones were divided into two groups, which were denoted "male" and "female." This understanding was connected with multiple material-semiotic interventions into human and nonhuman bodies: in midcentury studies that are now referred to as "classical," sex hormones were injected into pregnant rats, guinea pigs, and rabbits to investigate the impact on their offspring's physical and behavioral sex (Fausto-Sterling 2000; Roberts 2003a); humans diagnosed as sexually pathological (male homosexuals, for example) were given experimental hormonal "treatments" in attempts to "restore" "normal" sexual feelings and behaviors (Terry 1999); transsexuals wanting medical assistance to physically change their secondary sexual characteristics were offered experimental hormonal medications (Hausman 1995); and children born with indeterminate genitals were prescribed schedules of body-altering chemicals in connection with surgical interventions (Fausto-Sterling 2000). Through these experimental interventions and connected discursive argumentation in scientific and medical journals, sex hormones came to be figured as "messengers of sex" within a complex informational system that maintained sexual difference as dichotomous (Roberts 2002a). Male sex hormones, or androgens, came to be seen as productive of masculine bodies, brains, and behaviors, while the absence of these came to be seen as productive of femininity. Female hormones were understood to be responsible for secondary female sexual characteristics, and, to an indeterminate extent, to be antagonistic to maleness (Oudshoorn 1994; Roberts 2002a, 2003a).

The systems of biochemical feedback produced through these material interventions into the bodies of animals and humans gradually became standardized as scientific "facts." As such, they are represented in late twentieth-century physiology textbooks in simple line drawings, with hormonal pathways linking the neurological and reproductive systems.[4] The endocrine glands shown in these images are discrete entities, contained within the line diagram of the external skin of the adult human. Analyzing one such image from a standard endocrinology textbook, the feminist biologist Linda Birke points out that the diagram is full of empty white space. Organs float disconnected from each other, thus assisting in figuring the body itself as a free-floating, nonsocial entity (Birke 1999: 68–70). Such diagrams, and often the accompanying text as well, Birke argues, render the body an abstraction—scientific, mathematical, sealed.

Strangely, as Birke notes, these standard line drawings often depict a man whose body features ovaries as well as testicles. Although it would be difficult to argue that these images indicate the disappearance of sex—the images are often highly clichéd representations of white Western maleness and femaleness[5]—they do seem to suggest that the organs that "produce" sex are at least mobile, if not entirely detachable In any case, despite this notable and odd flexibility, the hormonal systems in these representations remain within individual bodies.

In late twentieth-century mainstream Western scientific representations, then, hormones are messengers within contained and abstracted systems of internal chemical feedback. Hormones participate in the "natural" production of sexual difference: androgens produce maleness, and the absence of androgens produces femaleness. The biochemical information systems producing sexual difference are internally coherent. It is this system that is now beginning to break down at the turn of the twenty-first century. This breakdown stems from two factors: first, developing understandings of the systemic effects of hormonal medication on bodies; and, second, contemporary concerns about the effects of hormones and chemicals that act like hormones (endocrine-disrupting chemicals) in the environment. In the next section, I outline these changes and critically discuss new forms of responsibility that are associated with them.

THE HORMONALLY SEXED BODY TODAY: FLUID ECOLOGIES

Medical Interventions

This model of the hormonally sexed body has always been connected to, and dependent on, particular forms of medical intervention. The central mode of intervention has been that of introducing human, animal, or synthetic hormones into the body in order to aid homeostasis when it is perceived as failing. Examples of this practice abound in the history of endocrinology, and they include the use of hormones to "treat" homosexuality, the hormonal treatment of aging in men as well as women (hormone-replacement therapy), and the use of DES to promote healthy pregnancies and minimize the risk of miscarriage. I have written about hormone-replacement therapy elsewhere, arguing that human, animal, and synthetic hormones have been used to maintain embodied aspects of dichotomized sexual difference (Roberts 2002b, 2003b). Others have described the use of DES from 1941 to 1971 (Seaman 2003; Seaman and Seaman 1977; Guisti, Iwamoto, and Hatch 1995). This prescription of synthetic estrogens was intended to rectify the drop in hormone levels that was experienced by some women in pregnancy, and which was thought to be linked to the risk of miscarriage. More than four million pregnant women were given large amounts of synthetic estrogen on the assumption that such a medical intervention into the hormonal aspects of pregnancy could only "improve" on inherent bodily processes (Seaman 2003: 46).

Each one of these interventions has proved problematic. In the case of hormonal "treatments" of homosexuality in the mid-twentieth century, many people suffered serious side effects from hormonal medication, while their supposedly problematic sexual desires remained unaltered. In the case of hormone-replacement therapy, it is becoming more and more widely accepted that the use of estrogens for the treatment of women's menopausal symptoms has potentially negative effects on reproductive and other organs. Indeed, the world's largest clinical trial of hormone-replacement therapy for women was prematurely halted in 2002 because of serious adverse affects experienced by women in the test group (Writing Group for the Women's Health Initiative Randomized Controlled Trial 2002). The case of DES, however, probably provides the most dramatic evidence of the negative impacts of the hormonal-intervention model. In the early 1970s, it was discovered that the teenage children of women who had taken DES during pregnancy were at increased risk of genital malformations, infertility, and rare forms of reproductive-tract cancers (Guisti, Iwamoto, and Hatch 1995; Seaman 2003). The women who took DES are also at increased risk of breast cancer and other cancers. More recently, animal evidence has indicated that the grandchildren of these women may also be at risk of reproductive-tract cancers (Newbold, Hanson, Jefferson, Bullock, Haseman, and McLachlan 1998). In many ways, then, it could be suggested that the model of intervention based on the idea of hormones as homeostatic systems, into which medicine can make quite simple, limited interventions, is now seriously contentious.

Endocrine-Disrupting Chemicals

These contemporary debates around forms of hormonal intervention and their impacts on sexed bodies take place alongside growing scientific and public attention to hormones and endocrine-disrupting chemicals in the environment (Roberts 2003a). In the contemporary scenario, as I will show, hormones are no longer contained within individual bodies. They do not circulate quietly along their own internal pathways; rather, they are understood to be everywhere, engulfing us in what is sometimes referred to as a toxic sea. This engulfment is often described as having a feminizing impact, as in the title of a popular book on endocrine-disrupting chemicals, *The Feminization of Nature* (Cadbury 1998). Such feminization is also characterized as an assault: the 1996 British television series on which Cadbury's book was based is titled *Assault on the Male*, and the best-selling writer Theo Colborn and associates describe endocrine-disrupting chemicals as "an assault on our children" (Colborn, Dumanoski, and Peterson 1996: 238).

Evidence of this engulfing assault comes in multiple forms, but decreases in human sperm counts and increases in human and nonhuman male genital malformations are the most thoroughly reported. For example, on March 17, 2002, several major British newspapers published a story about research undertaken by the British Department of Environment, Food and Rural Affairs and the Natural Environment Council. According to *The Independent*, this research showed that "male fish are developing female characteristics in rivers all over the country. In some stretches all the male fish have been feminised" (Lean and Sadler 2002a). The BBC quoted a scientist, Dr. Susan Jobling, who said that the research should be taken as a warning to humans: "The issue is not just about fish. Everything that we eat, put on our skin, throw down the drain, ends up in the sewage treatment works and ultimately in the river. So one could argue that we are actually living in a sea of oestrogen" ("River 'Pollution' Sparks Fertility Fears" 2002).

The Independent, making creative links with some U.S. Environmental Protection Agency research on sperm counts in humans and other mammals, came up with an additional story two days later, with a headline announcing that British men are "less fertile than hamsters" and suggesting that pollution "may be to blame" for the industrialized world's "collapse in sperm counts" (Lean and Sadler 2002b). This second story made the leap between the British research on fish and an increasing number of reports arguing that Western men's sperm counts were decreasing. Indeed, as the article's headline proclaims, today even small rodents make more sperm than men do ("proportionately" more, the text adds). This report suggests that British men, like their piscine counterparts, are at risk from estrogens excreted in the urine of contraceptive pill users. Because much British drinking water is collected downstream from sewage plants, Lean and Sadler (2002b) explain, men are imbibing unsafe levels of estrogen in the water they drink. Thus the reference to a fluid environment of estrogens—the "toxic sea"—is both metaphorical and literal.

Today then, hormonal pathways are no longer confined or local; instead, hormones and endocrine-disrupting chemicals actively flow across the globe. In the contemporary scenario, hormones make strange connections among animals, food, and humans, across vast distances. Another media report tells us that endocrine-disrupting chemicals released from factories in the southern hemisphere have been discovered in the fat of polar bears in the Arctic Circle (de Bendern 2002). This fat is consumed by Inuit children in Canada and Alaska, who are subsequently put at risk of developmental and reproductive problems, as are the polar bears themselves. Not only is the Inuits' food source under threat because of the reproductive impact of endocrine-disrupting chemicals on polar bears, their own children's health and reproductive futures may also suffer a similar fate if they continue to consume this staple food.

Unlike the twentieth-century hormones described earlier, contemporary hormones and endocrine-disrupting chemicals in the environment do not respect boundaries—of space (between countries or bodies), of species, or indeed of time. In contrast to the hormonal body depicted as an internally coherent line drawing, the contemporary hormonal body is "assaulted" by external chemicals. Endocrine-disrupting chemicals have an effect across the world's hemispheres, work through the bodies of fish and bears and other beings, and have an impact on generations by disrupting their reproductive capacities. Such chains of connection indicate that hormones can be understood as what John Urry—a sociologist developing the work of two science studies theorists, Annemarie Mol and John Law—calls "global fluids" (Urry 2003: 59–61). The actions of hormones within these systems are uneven, emergent, and unpredictable. Like other "global fluids" analyzed by Mol and Law, Urry, and others—anemias, brands, the Internet, money, traveling peoples—contemporary hormones and endocrine-disrupting chemicals travel along surprising pathways, making unpredictable escapes and "breaking free" of linear clock time. As "global fluids," contemporary hormones and endocrine-disrupting chemicals have impacts on sexed bodies that are unpredictable, difficult to assess and understand, unevenly experienced, and chronologically folded and enduring.

In her analysis of late twentieth-century discourses on the immune system, the anthropologist Emily Martin argues that "one of the central attributes of complex systems is that, unlike mechanical systems, they are never in equilibrium. Everything is in flux, continuously adjusting to change" (Martin 1994: 143–44). This contemporary understanding of systems as being in constant flux is applied not only to the biological body but also to organizational systems and processes and even to products themselves. In all of these, flexibility is "increasingly seen as intrinsically valuable" (ibid.: 149). In this sense, then, the constant and unpredictable flows of contemporary hormones form part of a wider picture, in which figurations of systems of numerous kinds (biological and social) are moving away from mechanical or fixed models and toward more flexible, ecological ones.

New Risks, New Responsibilities

Discourses on endocrine disruption describe new risks and configure new responsibilities for governments and NGOs, corporations and individuals. The dangers of these chemicals are, unsurprisingly, of greatest interest to NGOs focused on the environment, such as Greenpeace, Friends of the Earth, and the World Wildlife Fund for Nature (WWF). In 2005, an NGO consortium including these three groups mounted a serious lobbying effort around new European Union legislation known as REACH (Registration, Evaluation and Authorisation of Chemicals), regulating the production and use of industrial and agricultural chemicals (European Environmental Bureau et al. 2005). This campaign, like others before it and its North American equivalents, did not ultimately achieve its goals of tightening safety standards for hazardous substances (Friends of the Earth 2006). NGOs face considerable difficulties in proving their claims about endocrine disruption, in part because the science in this field is controversial, requiring methods of assessing the combined effects of minute amounts of chemicals (parts per trillion) on complex biological systems (Krimsky 2000; Solomon and Schettler 2000; Sharpe and Irvine 2004). For example, prestigious national research bodies—such as the National Research

Council (1999) in the United States, and the Royal Society (2000) in Britain—call for more research into the mechanisms of endocrine disruption, but fail to give wholehearted support to environmentalist NGOs' claims amid continuing technoscientific uncertainties.

Given the difficulties involved in getting governments to regulate endocrine disruptors, NGOs and other bodies recommend the so-called precautionary principle to citizens, asking them to take on new responsibilities for managing the as yet unknown effects of hazardous chemicals. These new responsibilities, I suggest, are similar to those discussed by Stacey (2000) in connection with cancer risks, and, like those, reflect the contemporary dispersal of the body into informational "bits."

Stacey analyzes the healthy body in terms similar to those used by Martin (1994), arguing that in late twentieth-century "self-health" or "alternative" medical discourses "the body is represented as a self-regulating system embedded within an ever-widening set of interconnected systems culminating in those of global proportions" (Stacey 2000: 129). The fact that the body is understood as a continual set of flows (Martin's "system in constant flux") means that health is a state that can always be achieved through adjustment of our individual responses to flows of all sorts: "If the body is no longer perceived to be a purely mechanical entity, but rather an informational network, then its openness to transformation becomes much more feasible and the obligation on individuals to monitor the harmonious balance of their interconnecting systems becomes a stronger imperative" (Stacey 2000: 131).

Stacey's work focuses on risks pertaining to cancer, and on individuals' responsibilities for maintaining their health before or after a (potential or actual) diagnosis of cancer (Stacey 1997, 2000). She also shows how discourses around responsibility for cancer particularly implicate women in their traditional role as caretakers of the next generation: "Women are thus called upon not only to be responsible for their own health and well-being, but to be responsible for the next generation's vulnerability to cancer" (Stacey 1997: 220; Stacey 2000: 131–40).

Fears about estrogens or endocrine-disrupting chemicals in the environment also create new responsibilities for individuals, and specifically for mothers. A British educational campaign developed in 2002 by the World Wildlife Fund for Nature and titled "Who Cares Where Toxic Chemicals End Up?" provides a good example. Published in popular magazines and on the World Wide Web, this campaign used images of fetuses, young children, and pregnant women to warn the public about the dangers that environmental toxins pose: "The womb should be the safest place on earth" the text of one campaign poster states, "but today our bodies are contaminated with over 300 man-made chemicals, to which our great grandparents were never exposed" (World Wildlife Fund for Nature 2002).[6] Such exposure, the poster urges, needs to be eliminated "so that the only thing we pass on to our children is our genes." In this campaign, environmental estrogens are figured not as a risk to women themselves but, through them, as a risk to their offspring. Women's bodies, in contradistinction to their "natural" function as housing "the womb . . . the safest place on earth," are positioned here as vectors for hormonal actions, not as their targets.[7] This angle has been picked up in newspaper articles on endocrine-disrupting chemicals. In a recent article in the *Guardian*, for example, Nicola Baird, a journalist, describes herself as "carrying a toxic time bomb," having discovered all the endocrine-disrupting and other pesticide-related chemicals present in her body (Baird 2002). In the *Independent on Sunday*, another journalist, Emma Cattell, encourages women who are worried about endocrine-disrupting chemicals to "make the biggest efforts" to protect their unborn children: "Finally, consider making the biggest efforts to cut exposure during pregnancy and when your children are young. Exposure in the womb to synthetic chemicals is considered [the] crucial factor in determining someone's long-term, adult health. Using 'natural' toiletries when you're pregnant and deciding your baby will be relatively 'cream-free' is infinitely more practical than [forgoing] cover-up for a lifetime" (Cattell 2003).

As vectors for hormonal risk, women are specifically targeted by educational campaigns aimed at encouraging consumers to minimize the impacts of endocrine-disrupting chemicals. A 2001 British campaign produced by the Friends of the Earth and the Natural Childbirth Trust, for example, provided mothers with advice on why and how they should avoid the "unacceptable threat to our health and the health of our families" posed by "risky chemicals." Printed on the back of a colorful fold-out height chart for children, the campaign depicts women using cosmetics and cleaning fluids and allowing babies to bathe in potentially toxic water while reaching for their potentially damaging plastic toys. Friends of the Earth also ran a publicity campaign in 2002 that involved transporting a seven-foot-high cuddly "Toxic Ted" to British shopping malls, in an effort to convey the message that things that seem safe for children may not be (Burley 2002: 25).[8] Through such campaigns, very local actions—bathing children, wearing cosmetics, buying toys—become choices linked to offsetting new globalized threats to the hormonally sexed body. As Stacey (2000: 132) argues, it is a requirement of contemporary citizenship to monitor and negotiate constant global threats and risks: "Global subjects are constituted through a sense of belonging which requires simultaneous monitoring of the body as an environment, and of the environment as embodied."[9]

In recent media coverage of the impact of endocrine-disrupting chemicals on sperm counts in Britain, men are also being encouraged to modify their behavior in order to attempt to increase their counts: cutting down on drinking, ceasing to smoke, and avoiding hot baths and tight underpants are often cited as potential changes. In such lists of advice, however, men are never expected to be responsible for the health of others; that is left up to women. Indeed, some articles trying to explain the decreasing sperm counts point the finger at men's mothers: low counts, they argue, may well be due to women's consumption of endocrine-disrupting chemicals while pregnant. Clearly, then, responsibility for monitoring the sexed hormonal body as an environment is gendered.

FEMINISM AND THE BIOLOGICAL BODY

My argument in this chapter is that the hormonally sexed body is currently undergoing a significant transformation. This transformation demands attention from feminists, not only because hormones remain a significant player in the production of sexed bodies but also because it is a transformation indicative of broader shifts in contemporary understandings of the biological body and in relations between the biological and the social, or in what, following Haraway (1997) and Rabinow (1992), could be called "technobiosociality."

I have argued that the modern sexed hormonal body, dating from the early to the late twentieth century, was (1) bounded, (2) able to achieve homeostasis through internal regulation, and (3) something that medicine assumed it could intervene in, without negative consequences, to assist in the maintenance of homeostasis.

Since the late twentieth century, this version of the modern hormonal body has become seriously contentious. In part this is because many of the major medical interventions based on this model are currently the subject of significant scientific, medical, and public debate. The debate encompasses concerns about the systemic impact of hormone-replacement therapy on menopausal women (the seriousness of which is indicated by the premature termination of the world's largest clinical trial of hormone-replacement therapy) and discussion of the findings of animal studies indicating that even the grandchildren of women who took DES may be at increased risk of reproductive-tract cancers. Hormonal interventions, it now appears, do not remain localized in particular regions of the body but instead have impacts at the molecular, genetic level. The impacts of hormonal medical interventions can have profound implications for subsequent generations.

Alongside developing concerns about the ongoing impacts of hormonal interventions into the sexed body, there is, as I have shown, growing debate about the impacts of hormones and endocrine-disrupting chemicals on the environment. The effects of these chemicals are seen to be globalized, diffuse, and unpredictable. These, too, are seriously contested.

Analyzing these two areas of the debate leads me to suggest that today's hormonally sexed body looks quite different from its earlier incarnations. The contemporary hormonally sexed body is (1) unbounded (geographically and chronologically) and affected by unexpected connections, (2) always under threat, as a consequence, from events and occurrences at the global level, and (3) something without intrinsic unity or coherence except as a nostalgic memory of our great grandparents' time (World Wildlife Fund for Nature 2002), when the hormonal body could achieve homeostasis.

I have argued that this new version of the hormonally sexed body produces new responsibilities for governments, corporations, and, what is most significant here, individuals. These individualized responsibilities, I suggest, are gendered in normative ways: the hormonally sexed body is now no longer a responsibility only for medicine but is also a project for modern women to engage with—to monitor, protect, and care for. Deciding whether or not to engage with hormonal medical interventions has become a complex responsibility for many millions of women today and is one that no one can take lightly (Roberts 2002b). Women are also asked to consider their bodies as vectors for hormones and endocrine-disrupting chemicals in the environment, and to change their everyday practices accordingly.

The point of this chapter is not to argue that the modern hormonal body has disappeared but rather to suggest that it is being seriously challenged. Both models—the new and the old—are in operation today. This fact in itself creates tension. This tension, I suggest, feeds into excessive demands on women to manage their own and others' hormonal bodies in the face of multiple uncertainties.

What are the implications of this argument for feminist theorizing about biological sexual difference and the body? In contemporary reconfigurations of the hormonally sexed body, female hormones (and chemicals acting like them) are separated from women's bodies and understood to flow dangerously along environmental circuits. Discourses around the dangers of what media reports often call "gender-bending chemicals" reify a version of sexual difference not only as something necessary (and therefore to be vigilantly protected) but also as something problematic (femininity removed from female bodies is figured as chaotic and dangerous). As I have argued, women are expected to become guardians of "natural" sexual difference in the face of breakdowns in modernist models of homeostatic, sealed bodies.

My aim is not to suggest that the health effects of environmental estrogens are insignificant or "made up"; on the contrary, I am convinced that they require serious attention. Such attention, however, would benefit enormously from a critical feminist approach to contemporary reconfigurations of material-semiotic biological bodies, and to "the fruitful collapse of metaphor and materiality that animates technoscience" (Haraway 1997: 97). We need to find ways to disentangle the threads (see Haraway, interviewed in chapter 3, this volume) of the implosions of metaphor and materiality that are inherent in these reconfigurations. Only in so doing can we begin to address the health and environmental consequences of endocrine-disrupting chemicals without reproducing negative views of femininity or situating women as the sole guardians of others' bodies.

The reconfiguration of the hormonally sexed body that I have outlined here is taking place in the broader context of changes in those biological discourses that Haraway (1997) calls "technobiopower." Hormonally sexed bodies in an era of technobiopower are understood to be flexible or fluid ecologies rather than sealed homeostatic systems. Thus contemporary hormonal bodies, as objects for theorizing, become more difficult to locate: sex hormones are "everywhere" and can be described as global fluids. As various chapters in this book demonstrate, this situation resonates with other aspects of contemporary biological embodiment (the genetic body, the cancerous body, the reproductive body). Feminist corporeal theory in general, then, and work on biological sexual difference in particular, must develop new methods of approaching the material-semiotic entities we call bodies.

NOTES

This chapter was initially written in 2003 as a paper for a workshop that was part of the "Missing Links" exchange among feminist scholars at Utrecht, Nijmegen, and Lancaster Universities. I would like to thank the participants in that workshop for their helpful comments and responses. For an extended version of this argument, see Roberts (2007).

1. Haraway's use of the phrase "life itself" is taken from Franklin (2000).

2. This is Claude Bernard's phrase.

3. *Merriam-Webster's Collegiate Dictionary*, 11th ed. (Springfield, Mass.: Merriam-Webster, Inc., 2004), 599.

4. To see examples of these images, type "endocrine system image" into an Internet search engine.

5. Birke (1999: 82) also talks about the ways in which whiteness, in such diagrams, is figured as the norm through the excessive

use of "blinding, antiseptic whiteness" in the images.

6. See also http://www.wwf.org.uk/Whocares/page3.asp (retrieved May 31, 2007).

7. I discuss WWF's use of a fetal image in detail elsewhere (Roberts 2003a: 203). As a number of feminist theorists have argued, images of the "living" fetus, such as this one, figure women's bodies in absentia as dangerous, even violent spaces (Duden 1993; Haraway 1997; Hartouni 1997; Franklin, Lury, and Stacey 2000: 36).

8. See also http://www.foe.co.uk/campaigns/safer_chemicals/press_for_change/toxic_ted/index.html (retrieved May 31, 2007).

9. Stacey's work focuses on health discourses that draw particularly on Eastern philosophies and understandings of the body. The discourses around the risks posed by hormones and endocrine-disrupting chemicals in the environment tend to be more prosaic and do not draw on universal "truths" in the same way as those analyzed by Stacey. Nevertheless, they do share an emphasis on individual responsibility for managing global-level risks at the local level.

5 ▪ Parenthood and Kinship in IVF for Humans and Animals

On Traveling Bits of Life in the Age of Genetics

AMADE M'CHAREK and GRIETJE KELLER

With the Human Genome Project initiative of the late 1980s, and the completion of the very first human genetic map, in 2000, genes became the traveling bits of life par excellence. They became the center of attention, intervention, and study in scientific practice but also an issue and a matter of concern in public discourse and in everyday life. Obviously, these bits of life travel between generations, with the passing on of genetic material from parents to offspring. Nevertheless, in the context of in vitro fertilization (IVF) this traffic may be less straightforward while still affecting configurations of parenthood or kinship.

As a technology, IVF itself does not stay put. It has become itself a "traveling technology." IVF is no longer called upon only in infertility cases but is at the heart of various other technologies, such as stem cell research, preimplantation genetic diagnosis (PGD), nucleus (ooplasm) transfer, and animal and cattle breeding. IVF thus plays a part in different scientific and medical practices.

In this chapter, we are interested in the question of what IVF technologies "do" in different practices. In particular, we pursue the traffic of genetic material and the role attributed to such bits of life in the conception of parenthood and kinship. We also contrast and discuss two quite different IVF cases: the practice of cell-nucleus transfer for human reproduction, and the practice of IVF in cattle breeding. It was the interesting parallels and differences between these two practices that prompted our choice to compare them in this chapter. They are obviously dissimilar in that they pertain to different species, but the cases discussed here are not altogether unrelated. In terms of technical options, bovine embryology is usually considered a testing ground for human IVF. Nevertheless, the discourses connected with the two practices are quite distinct. Issues of parenthood, gender, and even national identity are much more clearly articulated in cattle breeding, whereas they are rather muffled in the case of human IVF. The two cases discussed here allow us also to discuss issues of parenthood and gender as well as how individuals are linked through technology. To these ends, we analyze published material (whether in print or on the Internet) and scientific practice. We offer an ethnographic reading of this material in the sense that we take into account not only the worlds represented but also the ones mobilized and enacted through them. Technologies, as we will show, are at the heart of enacting (possible) worlds and possible links between individuals.[1]

IN VITRO FERTILIZATION: PROCEDURES AND RISKS

Because IVF is taken for granted in the practices that we discuss in this chapter, we shall briefly address the procedure and some of its effects. In the clinic, an IVF treatment starts with hormone therapy that causes the female reproductive system to produce up to twelve egg cells at the same time instead of only one per month (this is so-called superovulation). During this period, the Graafian follicles are monitored through blood tests and echography.[2] Once the egg cells have matured, and while they are still being monitored by echography, they are harvested through a hollow needle inserted into the uterus. The egg cells are placed in a Petri dish[3] containing a medium and are fertilized within a number of hours. After three to five days, one or two embryos are placed back into the uterus. Good "rest embryos," as they are called, are also set aside and frozen in case they have to be placed in the uterus at a later stage. If the initial attempt leads to a successful pregnancy, or if the woman does not wish to proceed with the treatment, the rest embryos will either be disposed of or, in some countries, used for scientific research.

Although reproductive technologies like IVF have become routine, a woman undergoing such a treatment has usually traveled a long hard path (Pasveer and Heesterbeek 2001; Cussins 1998; van der Ploeg 1998; Rapp 1999; Ginsburg and Rapp 1995). Once on the path of IVF, the woman consents to a procedure that extends well beyond the walls of the clinic, and whose success remains uncertain. In order for the treatment to be effective, arrangements and adaptations have to be made that involve care and monitoring of the woman's body around the clock, for long periods of time (Pasveer and Heesterbeek 2001).[4] As Pasveer and Heersterbeek argue, the success of IVF treatment is evaluated on the basis of its "take-home baby rate"; the complicated experiences of the women (and men) enduring the treatment are not the basis of evaluation. Moreover, as Irma van der Ploeg has shown, in clinical practice the trajectory of diagnosis, consisting of intrusive and risky tests performed on a woman's body, is set apart from so-called therapy; thus, "the concept of 'diagnosis' invokes images of mere observation, whereas actual intervention is reserved for 'therapy'" (van der Ploeg 1998: 62). In this way, clinical practice obliterates the laborious work of the woman, which is nevertheless required for the treatment's success. Diagnosis seems merely to concern reduction of uncertainty and "management" of the environment (that is, the female body) in which the future embryo will grow, whereas therapy is reserved for interventions in the laboratory proper, that is, in vitro fertilization of the egg cell (van der Ploeg 1998). A statement by the molecular biologist Lee Silver underlines this medical focus on the "therapy" and the consequent erasure of the female body: "The birth of Louise Joy Brown represented a singular moment in the history of humankind. Not because IVF provides a cure for infertility (and an imperfect one at that), but because it brings the human embryo out of the darkness of the womb and into the light of the laboratory day" (Silver 2000).

But this kind of unbridled enthusiasm about IVF, and its presentation as an "enlightened" technology rather than as a lived bodily achievement, may actually be at odds with clinical practice. Cussins (1998) conducted ethnographic research in fertility clinics and observed that the clinical environment is manipulated so as to resemble the dimness, warmth, and humidity of the womb. According to Cussins, doctors offer various explanations for why this womblike environment is better for the embryo. One such explanation involves risk reduction in the fertilization procedure. Practitioners in the clinics, quite aware of the fragile and erratic nature of the in vitro procedure, try to manage risk by resemblance: resemblance between the clinical environment and the "natural" environment of the womb.[5]

Throughout the 1980s and 1990s, use of IVF treatment increased dramatically. The medical indications for such treatment were enlarged, with the result that more ambiguous cases of infertility were referred to IVF clinics (Pasveer and Heesterbeek 2001; Kirejczyk, van Berkel, and Swierstra 2001). IVF is by now a well-established, routine technology. Despite its routine nature, however, the risks of intervention, both for the woman's body and for the embryo, are still quite high throughout the process of IVF treatment. Yet IVF, as a routine component of a package coined "reprogenetic technologies" (Parens and Juengst 2001; Silver 2000), has also become a technological intervention for reducing risk. In the cases discussed in this chapter, IVF is the very technology to be used in managing the risk: the risk of genetic defects in the future child, and economic risk in cattle breeding.[6]

MAKING PATIENTS, PARENTS, AND PARENTHOOD

IVF practices clearly illustrate that persons, parents, and parenthood are not givens; rather, they are achieved in process. Van der Ploeg (1998), focusing on male infertility, has analyzed the politics of representation in the scientific literature on IVF. Her study shows that although the female body undergoes direct medical intervention, this intervention is highly dependent on notions of the couple and the fetus. Both the couple and the fetus are represented as suffering patients who need good medical care. Hence the interventions in the female body. Van der Ploeg argues that the mechanisms contributing to the making of these two patients have also facilitated the further development of IVF.

Cussins's study, too, shows that in order for treatment to be successful, the "involuntarily childless" must play, actively and appropriately, the role of "patient couple." If the partners fail, they may risk their status as patient: "When the patient couple does not stay in control, remain civil, and, above all, manifest their stability as a couple, their public persona as an appropriate patient couple breaks down" (Cussins 1998: 73–74). In addition, alongside the categories of the couple and the fetus, clinical practice has introduced such categories as "pretty embryos" and "crude embryos." These are not just clinical assessments. They also help practitioners decide the fate of the embryos: to be transferred into the woman's uterus, to be frozen for future use, or to be discarded (Cussins 1998: 93–94). Thus the fact of an embryo's being "pretty" or "crude" makes quite a difference in IVF practices, a difference that leads to practical as well as ethical choices. Technologies, once put into practice, produce categories and sites of intervention and lead to particular configurations, both of humans and of objects of medical intervention.

It is quite clear that IVF technology entails the co-production of parents, fetuses, and treatments; nevertheless, it is striking that clinical discourse still tends to "naturalize" parenthood. In her research on IVF technologies and kinship, Strathern (1992a, 1992b) discusses the curious matter-of-factness about parenthood and procreation that is found in medical discourse. In terms of conceptualizing kinship, she argues, the new reproductive techniques leave little to be taken for granted. The social and the natural, the technical and the biological, are densely intertwined and require new modes of address: "The paradoxical outcome is that facilitating the process [through new reproductive technologies] does not automatically assist the making of parents. It assists the making of children" (Strathern 1992b: 21). In this discourse, parenthood, despite technical intervention, is either taken for granted or represented as a stable entity. In the remainder of this chapter, we discuss technologies and practices that either disturb common concepts of parenthood, parents, and offspring or produce them in particular versions. We will show that parenthood is less a natural category than an object of intervention and an outcome of particular practices.

CYTOPLASM TRANSFER

In May 2001, there were fifteen children born in the United States who carried genetic material stemming from two women and a man. These "genetically altered" children, as they were referred to in the media, were birthed through a technique called "cytoplasm transfer," also referred to as "ooplasm transfer."[7] As early as the 1990s, infertility researchers had started to apply this method on an experimental basis for rare cases of infertility, that is, cases where the embryo falls apart in the process of an IVF treatment. Although it is still not quite understood why the embryo falls apart, researchers assume that it has something to do with the cytoplasm of the egg cell, the jellylike substance surrounding the nucleus of the cell.[8] Scientists decided to look for donors for that specific part of the cell. Cytoplasm was extracted from donor egg cells and injected into the egg cell of the woman undergoing IVF treatment.

The question is how this procedure is connected to so-called genetic alteration. Traditionally, DNA has been viewed as a substance located in the nucleus of the cell: the chromosomes. Quite soon after the discovery of the double helix, it became clear that the same chemical structure was also found outside the nucleus: on the organelles, called "mitochondria." These are the so-called energy suppliers of the cell. This type of DNA has been termed "mitochondrial DNA," or "mtDNA." The peculiar thing about mtDNA is that it is passed on to the offspring through the mother only. The egg cell consists of a nucleus and the surrounding cytoplasm in which the mitochondria are located, whereas the sperm consists of just a nucleus. In genetic terms, males contribute nuclear DNA and females contribute both nuclear DNA and mtDNA. To bring this discussion back to the technique of cytoplasm transfer, inserting cytoplasm from a donor egg into the egg to be fertilized means passing the mtDNA of the donor on to the embryo. The embryo thus ends up with genetic material (mtDNA) stemming from two women: the intended mother and the donor. This fact explains headlines like "Can One Baby Have Two Mothers?" and "Scientists Say Stop Three-Parent Babies."[9]

Predictably, this new technique caused apprehension in the media as well as within the scientific community. This apprehension was due to two factors. First, the technique is in fact an instance of germ-line modification, which is prohibited by law in most countries. Second, scientists still know almost nothing about the interaction between mtDNA and nuclear DNA or about the effects of this interaction on the health of the future child. It is precisely because of its unknown but permanent effect on future generations that germ-line modification is prohibited by law. In the case of cytoplasm transfer, the mixture of the mtDNA contributed by both women affects all offspring. If the child is a male, he in turn will not be able to pass this mixture down, but a female child will transmit it through the maternal lineage.

CELL-NUCLEUS TRANSFER

While this international debate was taking place, the Dutch Health Council (2001) issued a report on a related technology: cell-nucleus transfer. Because the report takes up several of the issues we are discussing here, we will analyze how it deals with this traffic in bits of life in relation to parenthood.

The report, titled *Nuclear Transplantation in Cases of Mutations in Mitochondrial DNA*, deals with medical possibilities of nucleus transfer in cases of mtDNA disorders and discusses the ethical implications of this technique. Mutations in mtDNA genes can lead to a wide variety of disorders, such as blindness (Leber's optic atrophy), diabetes, and myopathy.[10] Since mtDNA is passed on through the mother only, an affected woman is bound to pass such mutations on to her children. An egg cell may contain up to a thousand copies of mtDNA. Although it is likely that not all copies carry the mutated genes, it is still quite difficult to estimate the risk that the future child will be affected by these mutations. Techniques like PGD cannot solve the problem. The molecular biologist David Rubenstein and his colleagues (Rubenstein et al. 1995) have published a protocol for a reproductive technique—in vitro ovum nuclear transplantation, or IVONT—aimed at overcoming these difficulties. The Dutch Health Council suggests that this technique will "allow couples in which the woman has a mutation in her mitochondrial DNA to have a healthy child 'of their own' if they so desire" (Dutch Health Council 2001: 13; author's translation). In that case, the couple undergoes IVF treatment in which a donor egg cell is used. As in cytoplasm transfer, both the intended mother and the donor undergo hormone treatment to produce simultaneous superovulation. A cell nucleus is then retrieved from one of the egg cells of the intended mother and transplanted into an enucleated egg cell of the donor (see fig. 5.1). After this procedure, fertilization is performed in vitro.[11]

Like cytoplasm transfer, IVONT results in an embryo that carries genetic mate-

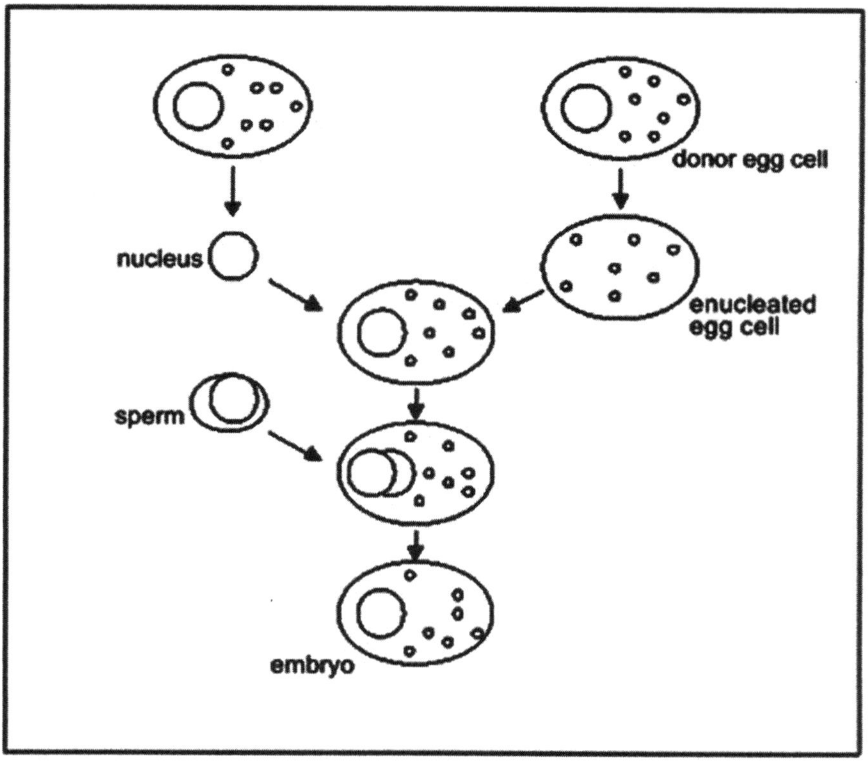

Fig. 5.1. Nucleus transfer. Dutch Health Council (2001)

rial from the intended parents (nuclear DNA from the father, and both nuclear DNA and mtDNA from the intended mother) as well as from the donor (only mtDNA). How does the Dutch Health Council deal with the issue of three parents? In its report, IVONT is viewed as a technique that helps a couple exercise the right to have a child that is healthy and genetically the partners' own. The report argues that IVONT should be treated as a kind of IVF treatment: "For those who are impaired by the mentioned illnesses, and whose wish to have a child is thus frustrated, the situation is no less severe than the involuntary childlessness of people who are enabled by society to undergo regular IVF treatment" (Dutch Health Council 2001: 41; author's translation).

"Regular IVF treatment" thus sets the standard for cell-nucleus transfer. People carrying mtDNA mutations are portrayed as suffering patients because the diseases are in many cases life-threatening, difficult to cure, and difficult to diagnose. IVONT is presented as a way of overcoming these problems, and as a technique badly needed by couples who want to "exercise" the "right" to a healthy child "of their own."

But the technique will not just produce a healthy child. It will also create a child with three genetic parents. To bypass this troubled parenthood, the Dutch Health Council's report trivializes the genetic contribution of the donor mother. The impact of her mtDNA on the individual is judged to be no more than that of bone marrow transplantation. Moreover, despite the intricate IVONT procedure, the report literally states that a child's personality is dependent on his or her nuclear DNA (Dutch Health Council 2001: 43). Similarly, the geneticist Gert-Jan van Ommen, discussing IVONT in the Dutch newspaper *Trouw*, says, "Mitochondrial DNA does not have anything to do with what makes a human being unique, his appearance and his mind. That is the work of *genetic material in the nucleus*, which these children will receive from their real mother" ("Een Kind van Twee Moeders?" 2001, emphasis added). Thus there seems to be a consensus that the identity of the child is formed by its DNA—to be more precise, by its nuclear DNA, not its mtDNA.[12]

The report of the Dutch Health Council (2001) emphasizes that the use of IVONT does not have to do merely with carrying a healthy child to term but also with having a genetic child of one's own. This emphasis is striking. What about the mtDNA of the second woman? The Dutch Health Council's emphasis downplays not the role of mtDNA—the initial worry, and the very reason for IVONT—but also the novel configurations of parenthood and lineage that IVONT helps produce. According to the Dutch Health Council, what makes a child the couple's own is the intended mother's contribution of nuclear DNA. It seems that the Dutch Health Council is keen to bring biological parenthood into line with social parenthood. Both are thus reduced to a question of nuclear DNA.

This approach, with its exclusive focus on nuclear DNA, denies and even erases the involvement of two women's bodies, that of the donor and that of the intended mother. For example, even though the Dutch Health Council (2001) deals with the ethical implications of IVONT, it glosses over the fact that the technique depends on the manipulation and synchronization of the bodies of the two women in such a way as to produce parallel, hormonally induced superovulation. The risks to these two women are not accounted for, by contrast with the ethical assessments that are made regarding the genetic risks to the future child, and by contrast with the moral justifications advanced for the technology in terms of the right to a child of one's own. The erasure of the contribution of mtDNA seems to go hand in hand with the downplaying of these risks. This erasure works as a rhetorical device that naturalizes parenthood and reproduces the couple as the natural origin of reproduction, thus depicting scientists and technology as simply "giving nature a helping hand" (Strathern 2002).

HOLLAND GENETICS

We have been arguing that different IVF practices may contribute to different configurations of parenthood. The second IVF practice that we will discuss here is one used in cattle breeding. Although it may seem odd, at first sight, to contrast human IVF practices to those used with animals, IVF actually does travel back and forth between the two sets of practices. For example, the IVONT technique was performed for the first time in bovines and only later applied to humans. Furthermore, research on in vitro maturation of egg cells is now being applied to bovines and is considered to be a future possibility for human IVF, especially as a substitution for the intense hormone therapy and superovulation that pose risks for the women involved. In this chapter, we will not go into the details of the traffic between the two sets of practices; we will focus, rather, on IVF in cattle breeding and its effect on kinship. For that purpose, we take a Dutch cattle-breeding company, Holland Genetics, as our example. The name of this company connects the long-standing Dutch tradition of cattle breeding to novel genetic technologies: along with tulips, the Dutch cow is an icon of Dutch national culture and landscape, and it is also deeply embedded in the (international) economy of Dutch farming; thus the linking of Holland with the field of genetics indicates that the legacy of the Dutch cow on the international market has not just a past and a present but also a future.

Holland Genetics is a subsidiary of the traditional Dutch cattle-breeding corporation CR-Delta. Since 1995, Holland Genetics has been using IVF techniques—or, as they are called in the field of cattle breeding, techniques of IVP (in vitro production)—to improve the quality of the livestock of CR-Delta's farmers and to produce a much higher number of elite bovines. IVP enabled Holland Genetics to introduce high-quality embryos, mainly from the United States, into its own livestock. In fact, IVP is a principal reason for the position of Holland Genetics as one of the world's top three cattle-breeding enterprises. It is now one of the leading companies in artificial insemination as well.

The company's success story takes on particular significance in connection with the Holland Genetics logo and the company's promotional activities. Just as human IVF practice often fashions the clinical environment to resemble the womb (Cussins

1998), Holland Genetics foregrounds features that make Dutch identity an unmistakable hallmark of the breeding environment. The clean environment of the laboratory is populated by such typically Dutch icons as posters of cows in a meadow, blue Delft tiles, tulips, windmills, and wooden shoes, and a Holland Genetics promotional video also exploits icons of Dutch identity in presenting the quality of the breeding program. The use of Dutch icons in the context of IVP practices has even been known to take on humorous characteristics in promotional images on the Holland Genetics website. For example, figure 5.2 represents spermatozoa as colorful tulips about to enter cows that have been designed in the colors of the Dutch national flag. Thus the high-tech environment of Holland Genetics is linked not to nature (the womb) but rather to national culture and tradition.

Let us now consider IVP in the Holland Genetics laboratory, where we will encounter the company's specific take on risk and risk management. The egg cells are immature when they are harvested from a cow's ovaries. This procedure is conducted every two weeks on "elite" cows when they are in calf, until the fourth month. The egg cells are then placed in a medium and put on a stove for maturation, which takes place over a period of twenty to twenty-four hours. After one day, the matured egg cells are ready to be fertilized with sperm cells from the Holland Genetics bank of high-quality sperm. To ensure that all the egg cells are inseminated with an equal number of sperm, the sperm cells are counted. Every single egg cell is fertilized with a sperm cell derived from a different top sire, a practice that enables the company to spread the risk (Brugh 2000). This spreading of risk is carried out with an eye to future revenues and the birthing of high-quality offspring for the company: if the embryos are carried to term (an outcome with a 50–50 chance of taking place), and if it is estimated that the calves will perform well in terms of milk production and other traits, then Holland Genetics has the right to purchase the best of them for its breeding program.

A Petri dish containing egg and sperm cells is then put on the stove. (The stoves are "the heart of the company," as a laboratory staff member told one of the authors.) The fertilization process is closely monitored over a period of eight days. At the conclusion of this procedure, the embryos are taken up in a straw and placed into portable stoves to be transported to various distribution centers, where farmers can purchase the embryos. The farmers inform the company when their cows are in the "right phase" to receive the embryos. The fertilized embryos are expensive by comparison with sperm (one embryo may cost between eight hundred and a thousand euros, whereas the price of a straw of sperm is about forty euros), but there is a considerable market for them. Holland Genetics exports embryos to countries all over the world.

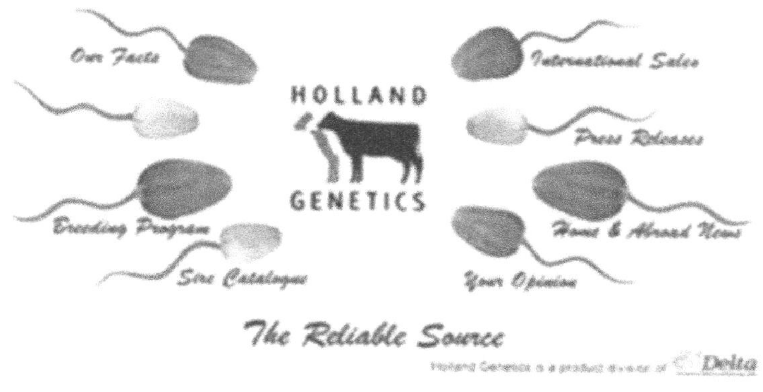

Fig. 5.2. A page from the website of Holland Genetics (2001)

BULLS WITH QUALITY UDDERS AND BEAUTIFUL LEGS AND FEET

When we consider the discourses of parenthood, something peculiar happens in the context of cattle breeding. Genealogies and records of pedigrees are pivotal in this practice, and such information, as well as information about the performance of individual cows, especially dairy cows, is laboriously compiled and recorded. Paternal and maternal lines are both important for the quality of the pedigree. Nevertheless, the information at Holland Genetics and similar companies seems to suggest that the father, the sire, is of much greater importance. Paternal descent seems to provide the most vital information about the performance of a future dairy cow. And it is odd information indeed. For example, the sire Kirby, we can read, "maintained his scores of 112 for frame and 110 for udder," and about the sire December and his daughter Eke we can read: "Eke is the perfect example of December. Beautiful, black, silky-skinned cow with super udders and great legs and feet." Similarly, the website of Holland Genetics contains elaborate lists of sires. These catalogs include information about the sires' pedigrees but also mentions liters of milk production, the percentages of fat in this milk, and so forth. In many cases a video is also provided on the website. The web user who clicks the icon next to the name of a sire and expects to have a glimpse of sire December, Don Juan, or Kirby is instead presented with images of dairy cows in a meadow or in a show. According to the specific qualities attributed to the sire, the camera diverts the viewer's eye to close-ups of magnificent udders or displays in a wider shot the "style and body" of the dairy cow and shows how she "paraded on great legs and feet."[13]

This puzzling practice of collapsing specific qualities of the male parent with those of the dairy cow not only presents an interesting case of gender bending but also points to a trivialization of the female parent's contribution. Nevertheless, in an interview that one of the authors conducted with the cattle breeders, they assured us that the female parent is equally important where the qualities of the offspring are concerned. How, then, can we explain the primacy and precedence of the male's contribution over the female's? A farmer named Kolff, a top breeder himself, answered this question by pointing to the pragmatics of the traffic in bodily materials. Some bits of life simply travel farther and more lightly than others. Egg cells are hard to store and distribute outside the body of a cow; sperm, by contrast, is easy to retrieve, maintain, and distribute, and so male qualities have moved to the center of the business. Moreover, a cow is usually in calf only once a year, whereas a sire can fertilize up to one hundred thousand cows in a single year. It is a question of numbers: size does matter.[14] The image shown in figure 5.2 unwittingly illustrates this fact: the spermatozoa are gigantic in relation to the cows.

Because it is easier to distribute bovine sperm cells across the globe, the sires' traits have gained more relevance for cattle breeders. Thus spermatozoa have become the means of controlling or improving the quality of the livestock and hence have become the terms of reference. The quantity and the distribution of sperm contribute both to the proliferation of male genetic traits and to the added value of sires for a breeding farm. From the perspective of a single pedigree, both the female parent and the male are equally important; in terms of livestock in general, however, a single male bovine has a much greater impact on the inherited qualities of the offspring than does any single female parent.

If we compare the practices of IVF for humans and animals, we see that parenthood in human IVF is represented as a naturalized relationship among the father, the mother, and the child, whereas parenthood and kinship in cattle breeding are represented, by virtue of tradition and pragmatics of the trade, as a relationship between the male parent and his multiple offspring—the livestock.

TIME IS MONEY: HOW DOES IVP AFFECT KINSHIP PATTERNS?

But there is still another side to the story of cattle breeding. Despite the numerical asymmetry of thousands of sperm cells versus one or two egg cells, IVP can actually provide a more active role for the maternal contribution. At Holland Genetics, the stock of high-quality sperm is now well established, and IVP is viewed as a means of complementing it with a stock of high-quality egg cells. With the IVP technique, a cow may produce twenty-five to thirty embryos a year. Those embryos will be carried to term by other cows. Therefore, Holland Genetics is breeding so-called bull dams that can provide egg cells for the embryos to be distributed. Since egg cells cannot be stored outside the body of the cow, the high-quality cows function as living banks of egg cells. Thus the very technology of IVP makes the genetic contribution of cows, in the form of embryos, more mobile in space and time, offering a means of controlling and improving the stock of farmers across the globe.

Moreover, IVP adds yet another value for farmers and individual cattle breeders. In the Netherlands, for example, there is a group of ten farmers involved in IVP. They produce embryos from their own stock, and the technical procedure is carried out at Holland Genetics. For these farmers, however, IVP is a means of producing more sires:

> Suppose you have a good sire and a good dam and you want to breed their pedigree, but the dam keeps conceiving females. It can take a while before you have a male offspring, a sire whose sperm you can sell. The advantage of IVP is that it takes a shorter time to have more progeny from the female line, and especially more sires. . . . Sperm is simply much easier to market.[15]

To assess what this means in economic terms, we must consider that one ejaculation from a top Holland Genetics sire is worth about fifty thousand euros (Brugh 2000). IVP, as a technique of retrieving many more egg cells from a dam, provides farmers with a highly profitable "condensation of time" (van der Ploeg 1998), which for individual farmers considerably increases the chances of acquiring a high-quality sire in addition to profits from marketing sperm.

Note that the cows deemed suitable for IVP are called "bull dams," sire mothers. Their odd name refers to the breeding practice of using a cow specifically for the production of sires. It is not that this sire-producing cow is incapable of producing female calves; Holland Genetics will, if requested, offer bull dam embryos of the female sex. Nevertheless, given the profitability of high-quality sires and the possibilities offered by IVP, it has become clear that even though technology provides means for a different configuration of kinship, the mobility and marketability of sperm are what hold the male parent-offspring kinship line firmly in place.

COMPARING IVF PRACTICES FOR HUMANS AND ANIMALS

So far we have presented two IVF practices and encountered diverse ways of enacting and representing parenthood. Here, we will discuss the differences between those two practices.

We started out by analyzing human IVF practices, where we encountered the clinic in the form of a womb. As we have seen, practitioners in fertility clinics try to manage risk by resemblance. One technique of enhancing the success of the treatment as well as the health of the future child is the manipulation of the space of the clinic so that it resembles the natural, dim, humid environment of the womb. By contrast, in bovine IVF (IVP), the laboratory space is made to resemble the Dutch national culture through the intertwinement of typical Dutch icons with laboratory work. In bovine IVF practice, there are different risks and stakes. Rather than the "take-home baby-rate," in cattle breeding the major risks and stakes are commercial. Dutchness seems to stand in for tradition and, as a consequence, for financial success in the business of cattle breeding. These two different ways of manipulating the environment point to very different kinds of risks. For human IVF practices, the risks involve the field of health and the success rate; for cattle breeding, the risks are primarily commercial.

The enactment and representation of parenthood are also quite different in the two IVF practices. Whereas human IVF is focused on the couple's desire or even right to have their own biological child, in cattle breeding this is turned upside down: it is important for a cow to have the right father. In human IVF, the focus of the treatment and the success rates are directed toward producing a healthy child. The parents and parenthood are naturalized and taken for granted. In bovine IVP, the qualities of the dairy cow are ascribed to the male parent, and his position is elevated above that of the maternal line.

Discussing IVONT, we showed how this technology contributes to novel configurations of parenthood, resulting in a child with three biological parents. In debates on IVONT technology, however, this new form of triangular parenthood is overlooked if not ignored. In order to naturalize parenthood to the familiar form of one father and one mother, nuclear DNA has to function as the signifier of real parenthood. This signification is carried out through the downplaying of mtDNA's (pivotal) role. Biological and social parenthood are brought in line through erasure of the bits of life contributed by the third "parent." Thus social fathers and mothers seem "by nature" to be the procreators of the child.

In bovine reproductive practices, on the contrary, the expedient naturalness of biological parenthood is not necessarily the term of reference. In cattle breeding, the male parent traditionally provides the term of reference for procreation. As we have seen, several factors have contributed to the proliferation of male parenthood in cattle breeding, such as the number of sperm cells versus the number of egg cells and the mobility and marketability of sperm. We have argued that this type of parenthood is not so much linked to individual offspring as to multiple offspring: the livestock. From our ethnographic analysis, it is clear that the materiality of the social practices in which reproductive technologies are applied has an impact on how (biological) parenthood is understood and represented.

Another difference we have traced in this chapter involves the role of the parents in human IVF and bovine IVP. Both practices have to do with improving future progeny, but the aims are medical (with IVONT, the child's health) versus commercial (higher milk production, for example). Since the potential parents become involved in IVONT because of genetic mutations, and since the intervention is directed toward preventing a genetic disorder in the future child, the health of that child is placed at center stage. One could say that these potential parents become responsible for their child's health through the very availability of the technology. In bovine IVP, the starting point is to produce high-quality sires and dams. The goal is to make their strong characteristics mobile, and therefore salable and profitable. Mobility is the means of improving the livestock of many different breeders at the same time. As we have shown, IVP is especially appreciated as a means of breeding high-quality sires in a much shorter time so that their sperm can be more quickly marketed. Rather than IVONT's burden of *responsibility*, IVP produces *opportunities*.

Technology is at the heart of both practices, but the practices' treatment of technology differs. In human IVF, technology remains somewhat hidden and serviceable. The technology is there to give nature a "helping hand" (Strathern 2002). With the technology veiled and concealed, both reproduction and parenthood can remain naturalized. In cattle breeding, the opposite is the case: technology is manifest, openly and everywhere. Holland Genetics stages the technology as a special feature of modern, advanced cattle breeding. It is marketed as a specific characteristic of a long tradition of good cattle breeding and as a guarantee of future success. Thus technology is celebrated. It is not so much an *addition* to a natural process as a *tradition* of cattle breeding.

Finally, a marked difference can be seen in the role that money plays in both practices. In human IVF, finance is disentangled from the actual intervention. (It should be borne in mind that in the Netherlands and most other European countries, IVF technologies and practices tend to be part and parcel of national health care and are therefore, for the most part, free of charge or covered by the individual citizen's health insurance.) At issue are patients in need of good care. In this context, the relevant category is not money but rather good medical service, whether it is provided by the state or by private insurance companies. The treatment can be discussed in terms of its success rate, but not in terms of profits or revenues. The costs of such treatments rarely figure in public discourse. And, once again, the situation is quite different in cattle breeding. Information about costs and revenues in IVP is openly shared. Price lists for sperm cells and embryos are available to farmers purchasing the products of Holland Genetics, as are estimates of future revenues. Cattle breeding is a commercial business aimed at marketing and selling cells, embryos, sires, and dams to potential breeders. In this practice, money and technology go hand in hand.

CONCLUSION

Technologies travel and find uses in different contexts. The two cases considered in this chapter show the effects of particular technologies on the practices in which they are made to work, but also how these practices and contexts in turn affect the technologies. Although reproductive technologies help shape practices in novel ways, they also become integrated into ongoing practice. We have investigated two IVF practices and shown, on the one hand, how technology comes to play a part in established forms of parenthood and kinship and, on the other hand, how technology enables new interventions and a novel traffic in bits of life. Even though we chose to compare two practices by focusing on their differences, the implicit resonance between the two is quite unsettling. What if we look at each practice from the point of view of the other? Such an exercise may reveal some of the more controversial issues and even taboos pertaining to reproductive technologies. For example, if human IVF is embedded in a discourse of health, how does this technology's diagnostic aspect contribute not only to the improvement of the child but also to that of the human "stock"? If IVF aims at reducing and managing risk, what types of risk are really at stake? Can we also think about risk in financial terms, or in terms of other types of assets? As many anthropologists have taught us, the self-evident configuration of parenthood in terms of one father and one mother is an ideological construction. How is this particular configuration linked up with Western ethnicity, and with eugenic or hygienic notions of the "healthy" family? Such considerations may lead us to question where the needs and desires of the individual (or of the couple, for that matter) stop and where those of the collective start. What enabling or prohibiting role does technology play in these issues? It is clear that such questions can and should be raised in cultural studies of technoscience. We hope that our consideration and contrasting of human and bovine IVF has helped to begin the articulation of some of these questions.

NOTES

We are grateful to Holland Genetics, and especially to the members of the Department of Research and Development , who allowed Grietje Keller to enter their laboratory and participate in meetings, and who supplied us with valuable information. We also thank Jan Kolff for additional information, and we thank Annemarie Mol, Mirjam Kohinor, Nicolette van Duursen, and Catherine Lord for their comments and generous suggestions.

1. One of the authors, Grietje Keller, has conducted research at Holland Genetics, the cattle-breeding company that we discuss. She visited the laboratory and conducted interviews with scientists as well as with some of the farmers involved.

2. An echograph is an image of an organ, or of any other part of the body, produced by ultrasound waves.

3. A Petri dish is a round glass or plastic dish with a cover. It is used to observe the growth of microscopic organisms.

4. For the women interviewed by Pasveer and Heesterbeek (2001), the duration of treatment varied between a few months and fifteen years.

5. Compare how Van Wagtendonk, a Holland Genetics engineer, talks about bovine IVF (cited in Brugh 2000): "At the moment there are various experiments going on in which the embryo is put in different growing media [instead of in a single medium for the entire process]. This better resembles the variability in the womb. . . . It is difficult to mimic nature. One should therefore keep the growing period as short as possible. An embryo should preferably be in the womb or in liquid nitrogen."

6. Various other "medical" technologies—for example, cloning techniques, stem cell research, and preimplantation genetic diagnosis (PGD)—are aimed at the reduction or management of genetic risk.

7. By May 2001, up to fifteen children had been born with the help of cytoplasm transfer. Sources: http://www.suit101.com/print_article.cmf/4866/80004; http://www.skfriends.com/can-baby-be-from-2mothers.htm (both retrieved Oct. 2003 but no longer accessible).

8. An explanation offered by Dr. Jacques Cohen uses the familiar "blame the mother" cliché: "The DNA in an egg undergoes chemical 'conditioning' before fertilization, so that early development may proceed properly. This woman lacked the ability to properly condition the DNA of her eggs." See http://www.txtwriter.com/Onscience/Articles/mtDNAtransfer.html (retrieved March 29, 2007).

9. Sources: http://www.suit101.com/print_article.cmf/4866/80004; http://www.skfriends.com/can-baby-be-from-2mothers.htm (both retrieved Oct. 2003 but no longer accessible).

10. Little is known about the prevalence

of mtDNA disorders. A recalculation based on a study in the north of England suggests that there are 1,600 patients in the Netherlands; see Centraal Bureau voor de Statistiek (1999a, 1999b).

11. This procedure has two variants. In the first, the fertilized egg cell is transferred into the enucleated donor egg cell. In the second, the fertilized egg cell is allowed to divide a number of times. A number of identical fertilized nuclei are retrieved and placed in several enucleated donor egg cells. This procedure is a cloning technology and therefore represents a case of reproductive embryo cloning. Thus it is in fact possible to produce genetically identical individuals. The Dutch Health Council (2001: 33) has understandable reservations about this variant.

12. For the bias toward nuclear DNA and the glancing over of mtDNA in genetics, see M'charek 2005.

13. This citation is from a short article on sire Kirby at http://www.hg.nl/news/highlights-bericht.jsp?id=2430 (retrieved March 29, 2007). The focus on looks and aesthetics is striking, but the emphasis on "great legs and feet" is actually specific to contemporary milking techniques: the shape of the legs is important for the milking machines, and the stability of the feet is important for the cow's ability to maintain a standing position. The more a cow stands on its feet, the more it can eat, and the more milk it will produce.

14. Jan Kolff, interviewed by Grietje Keller, March 12, 2004.

15. Ibid.

6 ▪ From Rambo Sperm to Egg Queens

Two Versions of Lennart Nilsson's Film on Human Reproduction

METTE BRYLD and NINA LYKKE

Within the last few decades, pictures of the reproductive process of sperm meeting egg, of embryonic cell divisions, and of fetuses in various sizes and positions have become a well-established part of popular culture. No one is any longer taken by surprise to see a picture of a thumb-sucking fetus contained inside the fetal membrane, or to witness the first cell divisions of an embryo on the television screen. Science documentary photos of the reproductive process, circulating persistently in books, magazines, and films, on TV, video, and DVD, and on the Internet, have transformed gametes and embryos into pop-culture icons with a life of their own.

In this chapter, we shall focus on that development, using two recent science documentary films on human reproduction as our analytical prism. The first, *The Miracle of Love* (2000), is a Swedish production, made by Lennart Nilsson and Bo G. Erikson Productions for Swedish TV and co-produced with a number of other European TV channels; a Japanese TV company; and the Public Broadcasting System (PBS), the noncommercial U.S. public television network, via *NOVA*, a documentary series that originates with WBGH, the PBS station in Boston. The second film, *Life's Greatest Miracle* (2001), is, surprisingly enough, *NOVA*'s own, different version of the Swedish film.

In other words, we will analyze two versions of the "same" film; for, even though the two documentaries have different titles, with "love" and "life" as their respective catchwords, they should nevertheless be considered variants rather than different films. The principal name behind both is that of Nilsson, the world-famous Swedish science photographer. A significant overlap of identical imagery from this "master photographer's" camera, presenting us in two instances with epoch-making pictures of the encounter between egg and sperm and of the growth of the fetus inside the womb, makes the Swedish *The Miracle of Love* and the U.S. *Life's Greatest Miracle* variants rather than two separate works.

By the 1960s, Lennart Nilsson had already earned international fame for his photo documentaries on the development of the fetus. When *Life* magazine ran a feature article showcasing his pictures of "life before birth" (Nilsson 1965), the eight million copies of that issue sold out in only four days.[1] The article was published simultaneously in several major European magazines, including *Stern*, *Paris Match*, and the *Sunday Times* magazine. Since then, numerous books (in particular, Nilsson [1966] 1993), photos, and science documentary films, featuring ever more technically advanced shots of the reproductive events in the penis, the testicles and, not least, the womb, have transformed the name of Lennart Nilsson into a brand. If there is one name that is almost synonymous with the visualization of the inner microworld of human reproduction, and with the twentieth-century transformation of the womb into a public stage, it is Nilsson's.

As for the relationship between Nilsson and the producers connected with *NOVA*, the documentary *The Miracle of Life* (1983) is especially important.[2] It played a major role not only in boosting Nilsson's career in the United States but also in establishing *NOVA* as the most watched documentary series on PBS. Although it was originally broadcast nearly twenty years ago, this film, according to the *NOVA* section of the PBS website,[3] is still rated the most popular *NOVA* episode of all time. To be designated the most popular *NOVA* episode of all time is no mean feat, considering that today *NOVA* is not only the most popular science series on U.S. television (and on the Web) but also the most watched science television series in the world, one seen in more than a hundred countries. The blockbuster status of *The Miracle of Life* may thus explain why *NOVA* promotes the newer film, *Life's Greatest Miracle*, as a remake of the 1983 *The Miracle of Life*, with no reference to *The Miracle of Love*, its 2000 Swedish precursor. It may also explain why it is one of only a handful of *NOVA* episodes that can be viewed free of charge online, even though it is also sold on video and DVD through the PBS online store.

The popularity and iconographic status of these images of human reproduction make it important to scrutinize their constructions of gender, sexuality, ethnicity, bodies, families, and subjectivity from a feminist perspective. Many feminist researchers (for example, Martin 1991; Stabile 1992; Duden 1993; Haraway 1997; Hartouni 1997; Franklin, Lury, and Stacey 2000) have criticized the narratives of reproductive science in general, and Nilsson's photos in particular, for reproducing gender stereotypes. For years, these narratives have continued to ascribe Rambolike qualities to the sperm and to confer personhood on the fetus while reducing the mother to a headless container.

By comparing the two versions of Nilsson's most recent film—the original Swedish version, *The Miracle of Love*, and *Life's Greatest Miracle*, produced specifically for the U.S. market—we aim to update the feminist discussion. It seems that the writer and producer for *NOVA*, Julia Cort, has listened to the critical voices, which have been much louder within Anglo-American feminism than in Scandinavia, where popular science and the field of science communication have not attracted much attention from feminists. At any rate, the Swedish version presents us with the traditional stereotypes, but the *NOVA* version of the same film, interestingly enough, has changed the story of egg, sperm, fetus, and parents-to-be in such a way that, at least to some extent, it rises to the challenge of the feminist critique.[4]

Nevertheless, it is likely that more than good intentions are at play in the U.S. version, which constructs gender, ethnicity, family, bodies, sexuality, and subjectivities in a more "politically correct" way than the Swedish one does. Whereas the Swedish film was produced in the context of state television and was influenced by its duty to "inform" and lecture in a "neutral" scientific-positivist way, the U.S. television industry's market orientation, with its insatiable demands for something "new" and "entertaining" to attract viewers, has left clear traces on the work of the writer, Julia Cort, and the editor, Dick Bartlett. *Life's Greatest Miracle* does not instruct by lecturing; rather, it informs by entertaining: it provides "edutainment." The view at *NOVA* is that a science documentary should be "as entertaining as it is informative."[5] Edutainment involves the interpellation of the figure of an "anybody and everybody" couple to match the film's target audience. The couple should be neither too rich nor too poor, not too distinctively ethnic, neither too young nor too old, neither too feminine nor too masculine, neither too hetero nor too homo (after all, we must be able to believe that the partners did make a child together). Ideally, the couple should be a multicultural hybrid that includes something multisocial, multiethnic, and multigendered—in other words, a perfectly average American couple, whose ideality lies in the partners' ability to erase differences and thereby signal the kind of liberal multiculturalism analyzed by Haraway (1997: 259ff.). In the editing, much thought was given to how this hybridity would be realized in the figures of identification that the film offers its viewers. Undoubtedly, the edutainment perspective was also on the minds of the Swedish producers, but it is much more integrated and considered in the U.S. version.

In order to illustrate our points regarding the possible effects of the feminist critique and the significance of the market orientation, we shall take a closer look at how the story of egg, sperm, embryos, fetuses, and parents is told in the two versions of the film. Our analysis is based on a close reading. We begin with a discussion of the similarities between the two versions and then proceed to the differences. First, however, we will briefly introduce the theoretical tradition into which our analysis is embedded: feminist cultural studies of technoscience.

FEMINIST CULTURAL STUDIES OF TECHNOSCIENCE: THE THEORETICAL FRAME OF THE ANALYSIS

Like the aforementioned feminist analyses of Nilsson's images and of popular science's representations of human reproduction in general, our analysis is based on the branch of research called "feminist science studies." More specifically, we define it as feminist cultural studies of technoscience. As Lykke explains in chapter 1 of the present volume, this subset of feminist science studies emerged from the intersection of cultural studies, science and technology studies, and feminist research.

Feminist science studies, as described in Lykke and Braidotti (1996), is a product of decades of feminist critiques of how gender is represented in the natural sciences and technoscience. These critiques generally have evolved in close association with social constructivist readings of both gender and science. Feminist science studies, as a highly interdisciplinary area that draws on very heterogeneous bodies of methodology, has also evolved in diverse directions, taking a wealth of perspectives on science, from its making to its culture, its representations, and its rhetoric. The field of feminist cultural studies of technoscience, as a subset of feminist science studies, developed in synergy with feminist cultural studies and its focus on popular culture (Franklin, Lury, and Stacey 1991; Thornham 2000)and has focused on popular science, science communication, and mediated representations of technoscience and the ways in which they articulate powerful cultural fantasies (van Dijck 1998; Franklin, Lury, and Stacey 2000).

As the popularity of Nilsson's imagery clearly demonstrates, popular science is as much a part of pop-cultural image machinery as are soap operas and romantic novels. Both kinds of popular culture give shape to iconographies, narratives, plot structures, and discourses of the cultural imaginary (Dawson 1994: 48)—that is, to the imagined worlds in which cultural communities mirror themselves, and on the basis of which they form their identities. There is a difference between pulp fiction and popular science in that the latter lays claim to the authority and "objectivity" of the natural sciences; nevertheless, feminist and social constructivist cultural analysts have shown that both genres can be treated as narratives, melodramas, romances, and action or adventure stories, and that both nourish as well as feed on the cultural imaginary. Seen from within a social constructivist frame of interpretation, the natural sciences as well as their pop-culture mediators can be understood as "story-telling practice" (Haraway 1989: 4–5).

This approach makes it possible to analyze (popular) science representations by means of methodologies that originate in film, media, and literary studies (Bryld and Lykke 2000). This is the type of approach we apply in the following close reading of the two variants of Nilsson's film. From the standpoint of genre, both films can be defined as popular science documentaries; but in a continuation of the feminist-ironic tradition that exists in feminist science studies (Haraway 1991a: 149; Martin 1991), they can also be interpreted as articulations of different types of love stories. To read these films ironically is one way to read them deliberately "out of context" (Haraway 1989: 377; Strathern 1987: 251ff.; Bryld and Lykke 2000: 39) so as to displace the claims of truth and objectivity implied in positivist science and, by extension, in the genre of the science documentary.

THE MIRACLE OF LOVE AND *LIFE'S GREATEST MIRACLE*: SIMILARITIES

These two film variants, like Nilsson's productions in general, can both be classified as "expository documentaries" (Kilborn and Izod 1997), that is, as belonging to the classic subgenre of documentaries that builds on the generic convention of "documentation," understood as a mimetic representation of a "real" reality. This type of science documentary is anchored in a positivist epistemology, and in its belief in the camera as a means of producing objective and true representations of a non-constructed reality. But this positivist illusion has been deconstructed by postmodern theories of photography as well as by feminist science studies, both of which direct attention to the dimensions of discursive construction in, respectively, photographs and the objects with which the natural sciences deal. Nevertheless, the contributions of these two fields have had little impact on the producers of science documentaries, who as a rule depend heavily on the mainstream of positivist natural science. For example, it is still the norm in science documentaries for an authoritative voice-over to comment on the objects represented in the film's images, in a language that pretends to scientific "neutrality," as if biomedical photography were a medium that could give us a totally transparent window on the world "out there"—or, in this case, the world "in there," inside the body. Nilsson's films are no exception to this rule. From his earliest productions to *The Miracle of Love* and *Life's*

Greatest Miracle, his most recent, all his films feature an authoritative narrator whose readings of the images are underpinned by the numerous science experts and institutions listed in the credits. With their names and reputations, they all vouch for the films' scientific qualities and objectivity.

The films are also very similar in their formal composition. The main story in a Nilsson film on human reproduction invariably follows a set pattern. The first act focuses on the encounter between egg and sperm. We follow the sperm cells from the time of ejaculation, and the egg cell from the time of ovulation, to the culmination—the coming together of sperm and egg. The second act deals with the development of the fertilized egg, from the first cell divisions until birth.

The film, acting as a frame for this report from a bodily microworld, which advanced medical equipment and sophisticated visualization technologies have made accessible to us, shows sequences that illustrate fertilization, pregnancy, and/or birth from the standard time-and-space perspectives of the viewers. In our two variants, this framing device constitutes a proper story with a narrative of its own. Each of these films, *The Miracle of Love* and *Life's Greatest Miracle*, presents us with the story of a heterosexual couple: the parents-to-be. This framing story creates a context for the microlevel narratives that underline the film's character of discursive construction. The generic conventions, and the narrator's scientifically authoritative interpretation, give the appearance of neutral, objective science to the microimagery of the events taking place in the testicles, the penis, the fallopian tubes, and the womb; by contrast, the framing story of the couple makes explicit the discourses at play.

In both film variants, the couple is represented as heterosexual partners whose mutual love leads them to spontaneous fulfillment of their genetic programming for reproduction. The representation of the couples differs in the two films, however. Whereas the Swedish version chooses to focus on a romantic-mythic loving couple, the U.S. film presents realistic, everyday conjugal happiness. The common denominators are still love, happiness, and heteronormativity (although there are certain normativity fractures in the American variant, as we shall see). In both films, this hegemonic normativity is displayed by the figuration of two partners who are meant, in their own way, to represent "anybody and everybody," and whose appearance and ways of acting are naturalized by the microlevel story. The latter unambiguously thematizes biological determinism and genetic essentialism.

Yet another similarity between the films is an obvious hierarchy between the macro- and microlevel stories. The comments of the authoritative voice-over thus leave us in no doubt that the "objectively scientific" report on the microlevel contains the proper explanations of what is going on, and this message is very clear in both variants.

As a prelude, before the introduction of the couple, the U.S. film bluntly maintains that what is really at stake is occurring at the microlevel, "under the skin," where DNA and genes rule. While the camera pans over a vast, more than half-naked crowd on a beach, the narrator lectures us on "the urge to procreate":

> You might think all the people on this beach are just working on their suntans. But beneath all that sunscreen, under the skin, there's a frenzy of activity. Without even thinking about it, almost all adults here are busy trying to reproduce. They can't help themselves. The urge to procreate is a fundamental part of life, not just for us but for all life.

In the less edutainment-conscious Swedish version, the hierarchy between the framing story and the microlevel story is even stronger. The glamorous framing story seems to be merely a decoration, the sole purpose of which is to arouse the interest of a sleepy TV audience in "proper" scientific messages on "the urge to procreate." The romantic couple, the film wants to convince us, are "in reality" driven by their genetic programming.

Basically, the rhetoric of both films is firmly rooted in the discourses of sociobiology, which has been very influential in the biological sciences ever since Wilson (1975) made his mark. Sociobiology is characterized by a neo-Darwinian combination of genetic essentialism and evolutionary utilitarianism, which stresses that the most essential task all living beings, one they are programmed to fulfill, is to pass on their genes.

THE MIRACLE OF LOVE AND *LIFE'S GREATEST MIRACLE*: DIFFERENCES

Although there are clear similarities between these two versions of Nilsson's film, the differences are even more conspicuous. This is true whether we look at the story of the couple or shift our attention to the choreography of the gametes inside the female body. At both the macrolevel and the microlevel, the enunciations are very different. Let us look at the differences between the two framing stories before we turn to the microlevel.

The Framing Stories

In the original Swedish version of Nilsson's film, the story of the happy heterosexual couple is endowed with universal, romantic-mythic, almost religious dimensions (Lykke and Bryld 2002). The film opens with the image of a fetus whose hands move toward each other while the narrator, with almost religious pathos, tells the creation narrative of the child: the story of "one of the miracles of our life on Earth." Then the picture of the fetal hands dissolves into a wedding image of the bridegroom placing the ring on the bride's finger, and we are told that the creation of a child "often begins . . . like this: two people want to share their life with each other and hope that they will have children together."

After this bombastic prelude to a romantic-religious narrative, a series of short sequences presents the absolute highlights of the myth of the universal love-couple: the prelude to the partners' intercourse, which itself romantically takes place on a beach; their first shared joy at "seeing" the fetus in an ultrasound image; and then, at the hospital, the birth, which goes very easily, painlessly, and "naturally." The final scene presents the nursing mother and the suckling child in a green garden, and, as if to emphasize the many high-tech photographic "scoops" on which the film prides itself, the film suddenly takes us, the viewers, inside the baby's mouth, where we can watch the milk spurt from the maternal breast, almost as if we ourselves have been put to the breast of the white Swedish nature-mother.

The Swedish couple's mythic-iconographic status is underlined by the partners' namelessness, by their lack of subjective perspectives or voices (they never speak), and also by the fact that their lives are not in any way contextualized. Moreover, the roles of He and She are played by two actors so well known in Sweden that Swedish viewers, at least, are fully aware that what they see on the TV screen is only staged role-playing that, for all its glamour, might just as well be a soap ad, for cleanliness, whiteness, and immaculateness are precisely the qualities that the "natural" and happy couple radiates. What the Swedish film wants to show us is, in other words, an image of the abstract human being or, rather, a representation of the human that is ethnocentrically understood as a white, well-off young heterosexual couple, where the two "halves," She and He, complement each other in accordance with bourgeois and phallocentric norms. Their gender identities are clear and well defined. He is the active one, who takes the initiative. She is passive and receptive. We see Him put the ring on Her finger, and it is He who undresses Her in the prelude to their intercourse on the beach.

In contrast to the Swedish pathos and glamour, the U.S. version of the film presents us, as already noted, with everyday, and down-to-earth conjugal happiness. This framing story features two multicultural partners with names and voices of their own (there is no voice-over in these segments). They are the Anglo-American Melinda Tate Iruegas and the "Chicano" Sergio Iruegas. They are depicted in realistic documentary style, with emphasis on a phenomenological, experience-centered perspective. Also in contrast to the Swedish version, in which intercourse on the day of the wedding automatically leads to the "miracle" of a love-child, Melinda and Sergio have been together for some time before they feel ready to have children. Moreover, the fact that pregnancy can be controlled is strongly suggested by Melinda's words: "We weren't being as careful as we should have been." Thus the pregnancy is partly planned and partly the surprise implied by the idea of a love-child. In further contrast to the glamorous Swedes, Melinda's and Sergio's social status is vaguely connected with that of the (lower) middle class, and, most remarkably, their gender identities are conspicuously blurred.

Melinda is a bit taller than Sergio and seems in all respects to be the head of the family. Sergio clearly looks up to her and lets her take the lead. Thus, whereas Melinda is depicted as having "masculine" authority in the relationship, Sergio is both psychically and physically the soft and "feminine" aspect. When he is walking hand in hand with the pregnant Melinda, the incipient breasts of his chubby body are seen to swing under his T-shirt. Moreover, his being almost as pregnant as his wife is voiced in a story about how he used to get "really nauseous and upset" during the pregnancy, just like Melinda, and sometimes even "physically ill." Of course, this story may serve as an example of how pregnancy is currently ascribed not to the mother but to the couple (see Kirejczyk 1994). In the context of the microlevel story, however—which, as we shall see, reverses the traditional phallocentric depiction of the roles of egg and sperm—this blurring of gender boundaries acquires additional meaning. When gender is not "done" in the "normal" way, heteronormativity is fractured.

The phenomenological perspective of the U.S. version lets us get close to the couple. We follow Melinda and Sergio as they talk to each other or to the camera on pregnancy-related topics, in a relaxed, everyday language that makes it easy for viewers to identify with them. In a pedagogical way, we are presented with the good experiences and minor problems presumed to be typical of a couple expecting a first baby. Initially, we witness a conversation about family resemblances, in which a photo album inspires a discussion of what the future child will look like. We are then confronted with the couple's emotions and reactions to the pregnancy as it progresses chronologically. Eventually, we get a realistic forewarning of birth pains as Melinda relates an anxiety attack she had when, on a visit to the toilet, she was struck by the thought of how tiny the vaginal opening is compared to the child that has to come out of it. Finally, we witness the birth, where not only Sergio but also both of Melinda's parents are present. Just as the story began with family photos, it ends with a reaffirmation of the family in the context of the newly fledged parents and, indirectly, that of the image of the United States as one big multicultural family (see also Collins 1998; Haraway 1997: 213–65).

The Story at the Microlevel

If the U.S. producers endeavored to make the framing story more "politically correct" and less old-fashioned in its presentation of gender, family, sexuality, body, subjectivity, and ethnicity, this is no less true when it comes to the film's comments on the microimagery. Here, the two versions of the film differ even more radically.

First and foremost, the U.S. version of the first act (the coming together of egg and sperm) breaks decisively with an incredibly tenacious biomedical rhetoric on human reproduction, one wittily criticized by Martin (1991) in her famous article on how science traditionally constructs the encounter between the two gametes. All Nilsson productions are inscribed in precisely this traditional phallocentric master narrative, which portrays sperm as Rambolike supermen (Stabile 1992) on an adventurous, dangerous quest for the Egg, who like a fairy-tale princess passively waits for her prince to come. Although there are minor discursive updates in *The Miracle of Love*, the plot as a whole is still firmly based on the heroic journey of the "sperm armada" with its "power plants" (the male gametes). All activity and drama are rhetorically and narratively ascribed to the sperm. By contrast, the egg is described passively (it is "transported" from the ovary) and in functional language ("the fallopian tube now tries to get the egg in"). This kind of language differs markedly from the militarily heroizing metaphors lavished on the sperm cells ("armada," "advance," "phenomenal swimmers," "struggle for survival," "the winner").

Against this background of traditionally phallocentric discourses on the encounter between egg and sperm, it is significant (perhaps even a discursive "miracle") that the roles in the U.S. version, *Life's Greatest Miracle*, have been switched, more or less. Here, it is not the sperm but the egg and the woman's body that run the show. Although the photographic shots in *Life's Greatest Miracle* are predominantly the same as in *The Miracle of Love*, the voice-over constructs the story in a radically different way. Both grammar and plot structure are now focused on the female elements as subjects of the process. They are the ones in total command of the sperm's journey as they "guide," "propel," "alter," "order," and "draw the entire contents of the sperm inside." Should the sperm show any sign of activity—such as, for instance, swimming through the fallopian tube—the narrator instantly informs us that these Rambolike swimmers will not have much chance of reaching the egg. Conversely, it will be the "slowpokes," caught up in the cilia lining the fallopian tube, and only "gradually released," that are likely to get "a date" with the egg. In contrast to the sperm that race along, we are told, the slow ones have a better chance of being altered by the woman's body and later being accepted by the egg. It is also emphasized that the egg will allow the sperm in only if the egg and its "picky chaperones" of "support cells" permit the sperm's entrance. "Brute force alone," the voice-over maintains, will not do the trick. Thus it is the matriarchally described egg that sets the agenda, as a reflection of how Melinda is presented in the framing story, as head of the family.

Yet another difference in the microlevel story separates the two versions: DNA and its agency play a much more prominent role in *Life's Greatest Miracle*. Here, rhetorical echoes of the biologist Richard Dawkins's theory of "the selfish gene" (Dawkins 1976), which sees the gene rather than the organism as a prime mover in evolution, resonate more forcefully than in the Swedish version. Even though the Egg Queen is cast in a crucial role in *Life's Greatest Miracle*, it is DNA, in itself not sex-specific, that has overall agency in the U.S. film's creation narrative. As already mentioned, we are informed very early in the course of events that "the urge to procreate" is situated in the body's DNA, that is, in the "molecule that carries our genes." Thus, finally, it is neither the female nor the male organism that acts as the procreational prime mover; it is DNA, common to all life. Undertones of a discourse of democratic equality are therefore at play, further setting *Life's Greatest Miracle* apart from *The Miracle of Love*, in which the role of DNA is not emphasized nearly as much. One reason for this difference is probably a purely technical-aesthetic one: the explanations of DNA offered by the U.S. film require computer-animated illustrations, and computer animation is reduced to a minimum in the Swedish version, which openly celebrates the unique craft of Nilsson, "the master photographer."[6] But the lack of focus on the overall agency of non-sex-specific DNA also strengthens the phallocentric perspective in the Swedish version. The basic agency in *The Miracle of Love*'s procreation story is concentrated in the sperm cells, as a reflection of the phallocentric depiction of sexuality at the macrolevel, where He is the active subject.

The last difference we shall mention deals with the representation of the mother and the mother's body, and with the connection between the framing story and the second act of the microlevel story. In all Nilsson's films, the second-act highlights consist of close-ups of the growing fetus. The fetuses are shot to look like "space creatures," floating weightlessly in a cosmic-looking setting (Duden 1993) that has very little resemblance to the walls of the womb. The reason why the womb is missing is very simple: the material for most of Nilsson's pictures is aborted embryos and fetuses that only appear to be portrayed alive because of the context. A few sequences do display living fetuses filmed in utero by means of special techniques, but not even here do we see much of the womb itself. As feminist critics have pointed out (Martin 1991; Stabile 1992; Duden 1993; Haraway 1997; Hartouni 1997; Franklin, Lury, and Stacey 2000), the mother is presented as a mere container, and even so we are barely permitted to see her. The focus and center of the microlevel story are clearly the fetus.

Whereas *The Miracle of Love* reproduces this traditional visual rhetoric, *Life's Greatest Miracle* breaks with it, not so radically as in the egg-meets-sperm figure but in a way that is still worth noticing. To begin with, the number of fetal pictures is reduced in the U.S. version, in favor of the framing story. Moreover, the phenomenological perspective of the framing story serves to represent the objectified mother's body on the microlevel in the context of Melinda's experiences during pregnancy. Finally, these two changes work together to tone down the contrast between a personalized, superactive fetus and a passive, objectified mother.

CONCLUSION

When we suggest that the differences between the two versions may be due in part to the U.S. scriptwriter's awareness of feminist critiques, it is because *Life's Greatest Miracle* addresses some of the main points made by feminist critics. The ethnically white-white couple in the Swedish version, with its patriarchal gender roles, has been ousted and updated in the U.S. version so that neither gender nor ethnicity is "done" in the traditional, (hetero)normative way. The same goes for the myth of Rambo sperm and the depiction of the mother as nothing more than a headless container for the personalized fetus.

In this context, it is interesting to note that *Life's Greatest Miracle* consciously emphasizes the rhetorical break with the outdated phallocentric sperm story, which the narrator explicitly dismisses: "Nothing could be further from the truth." Nevertheless, true to its conventional scientific outlook, the film presents this rhetorical shift as an expression of scientific advances, not as an outcome of negotiations with feminism or with viewers' perspectives.

It is clear, however, that more than scientific advances are at stake. It is indeed true that newer scientific theories of the reproductive process have established the egg, the vagina, and the fallopian tubes as more significant agents in the process than is the rather weak propulsion of the sperm tails. But, as noted by feminist researchers, these theories have had remarkably little impact on the way in which the sperm-meets-egg story has been told for years. The myth of Rambo sperm has been repeated tenaciously, despite scientific evidence that serves to undermine it (Martin 1991; Spanier 1995: 24). Moreover, these newer theories concerning the active role of the egg, and of the woman's body, were definitely around when *The Miracle of Love* was produced, in the late 1990s. Therefore, to ascribe the discursive reconstruction of the fertilization narrative in *Life's Greatest Miracle* to a sudden leap forward in knowledge would be too simplistic. What we suggest instead is that the reversal of roles—the rhetorical shift from Rambo sperm to ruling Egg Queen—should also be understood as part of a marketing strategy, as a way of trying to catch the attention of an audience that has grown tired of the sexually stereotyped tale of Rambo sperm and a passively waiting Snow White of an egg. At any rate, it is obvious that *NOVA*'s public relations material proudly advertises the discursive shift:

> Among the stunning new sequences shot by Nilsson is the incredible voyage of the sperm toward the egg. Sperm are often portrayed as brave warriors forging their way through hostile terrain, racing to overcome impossible odds, where the fastest and most powerful will conquer all and vanquish the egg. Nothing could be further from the truth. The journey of the sperm is controlled to a great extent not by the sperm itself but by the woman [that is, the woman's body, including the egg].[7]

We suggest that the efforts of WGBH and *NOVA* to develop marketing strategies for this kind of film may have been influenced by an interest in listening to feminists, and to women viewers in general. The schoolteachers or mothers-to-be who are likely to purchase the video or watch the film online, and eventually give their support to the Public Broadcasting Service, are by and large independent, wage-earning women. Whether feminists or not, they make up a large group of consumers who are looking for a renewal in images of women and in stereotypical gender representations. This is a need that the film industry and market-oriented television in general started to trace and explore rather a long time ago—from the 1991 film *Thelma and Louise* and the femi-crime genre to such TV series as *Ally McBeal*—but a need that nevertheless seems to have had difficulty breaking through to the producers of the specialist genre of science documentaries. Perhaps a new trend is in formation here, as possibly indicated by the conscious promotion of the egg-meets-sperm story in *Life's Greatest Miracle*, as an alternative to the tale of the heroic quest of the Rambo sperm.

Nevertheless, as we evaluate the potential for renewal in this possibly upcoming trend in science documentaries on human reproduction, we must stress that even though *Life's Greatest Miracle* represents a significant and welcome break with what is both a phallocentric and an ethnocentric tradition, it is not without its own problems. Let us end by briefly summarizing what we see as the most important ones.

First, reductionism and genetic essentialism are as much a part of *Life's Greatest Miracle* as of *The Miracle of Love*, the Swedish original. Human subjectivity is basically presented as a question of genetic programming—"the urge to procreate." Second, neither version of the film questions the expository documentary genre's traditional positivist claims to tell a true and objective story of the microlevel events—their claim, that is, to act as a transparent window onto a biological reality "out there." Epistemologically, both versions sustain the "god-trick" (Haraway 1991b: 189) of modern science, thereby underpinning the reductionist and essentialist messages. Third, against the background of the overall genetic essentialism of the two film variants, the reconstructed egg-meets-sperm story in *Life's Greatest Miracle* ends up being a mere reversal rather than a displacement of fixed, heteronormative sex/gender hierarchies. Fourth, and finally, the liberal multiculturalist discourse inscribed in *Life's Greatest Miracle* by its framing story can be criticized as a glossing over of power differentials and mechanisms of social exclusion, which obviously have a big impact on the conditions of family building and reproduction. As stressed by Collins (1998), the liberal multicultural version of the universal or all-American couple is a nationalist construct that does not evade the trap of ethnocentrism.

Thus, even if, from a feminist perspective, NOVA's version of the story of egg, sperm, embryo, fetus, motherhood, and fatherhood seems more promising, there is still much to be done. Critical feminist interventions and radically renewed narratives are much needed in popular representations like the ones discussed here.

NOTES

1. See www.albertbonniersforlag.com/900/900.asp (retrieved March 12, 2007).

2. The widely acclaimed Nilsson-*NOVA* documentary *The Miracle of Life* is more or less identical to the Swedish original, *The Saga of Life* (1982). By contrast with the newer production, *The Miracle of Love*, this film has no romantic-religious framing story.

3. See http://www.pbs.org/wgbh/nova (retrieved March 12, 2007).

4. The U.S. version was awarded an Emmy for writing. But Julia Cort also co-

authored the Swedish version with Bo G. Erikson. A transcript of *Life's Greatest Miracle* can be downloaded from the PBS website; see www.pbs.org/wgbh/nova/transcripts/2816 miracle.html (retrieved March 12, 2007).

5. See www.pbs.org/wgbh/nova/about/appr.html (retrieved March 12, 2007).

6. According to Erikson, executive producer of *The Miracle of Love*, the film even "stands as a protest and alternative to computer-manipulated projects made in the [name] of science[;] perhaps it's the last of its kind." Source: http://input2000.cbc.ca/programs/44.html (retrieved Jan. 2003 but no longer accessible).

7. Source: www.ptvpromo.org/program info/109/nova_lifes_miracle.html (retrieved Jan. 2003 but no longer accessible).

7 ▪ Screening the Gene

Hollywood Cinema and the Genetic Imaginary

JACKIE STACEY

In a short article titled "Sameness Is All," the psychoanalyst and cultural critic Adam Phillips offers an account of a conceptual confusion that became evident in one of his clinical cases. He tells the story of a child who had been referred to him for school phobia, which had started a year after her younger sister was born. Phillips (1998: 89–90) gives the following account of their therapeutic exchange:

> In her second session [she told me] that when she grew up she was "going to do clothing." I said, "Make clothes for people?" and she said "No, no, clothing . . . you know, when you make everyone wear the same uniform, like the headmistress does . . . we learnt about it in biology." I said, "If everyone wears the same uniform, no one's special." She thought about this for a bit and then said, "Yes, no one's special, but everyone's safe." . . . I was thinking then, though couldn't find a way of saying it, that if everyone was the same, there would be no envy; but she interrupted my thoughts by saying, "The teacher told us that when you do clothing you don't need a mummy and a daddy, you just need a scientist. A man . . . it's like twins. All the babies are the same." . . . I said, "If your sister was exactly the same as you, maybe you could go to school," and she said "Yes," with some relish, "I could be at home and at school at the same time . . . everything!"

This extraordinary story of confusion between cloning and clothing encapsulates perfectly a tension I wish to explore in this chapter: the discrepancy between the promise of what scientists can guarantee and what outward appearances can achieve, between biological and cultural design, between the reproduction of embodied identity and the production of image. This child's fantasy offers a resolution to the problem of sibling rivalry and parental control by positing a male scientist at the head of a system in which similarity guarantees safety (the absence of envy), and duplication overcomes the problem of displacement by a younger sibling (simultaneous presence in two different places): it seems to allow the child to have "everything" she wants all at once, to fulfill all her wishes simultaneously. The threatening potential for reproducing sameness at the biological level (cloning) is transposed onto the production of sameness at the cultural level (uniformed clothing).

The conflation of biological and cultural designs in this story raises the question of how identities are encoded and decoded. This in turn points to what we might in shorthand refer to as "the dialogics of identity": the ways in which identities are not simply constituted, embodied, or lived out by subjects themselves but are also always read by others.[1] This dialogic process of identity production requires multiple cultural competencies in order for an interpretative exchange to be achieved between subjects, or between subjects and institutions. But if we accept that identity is a process rather than a product, a dialogic effect rather than a personal possession, then genetic engineering and cloning complicate this picture in some interesting ways. On the one hand, genetics may appear at first to offer an anchor to the instabilities of identity formation in this postmodern world: genetics promise predictability, control, order, and security through new techniques of screening, selection, preimplantation diagnosis, and cloning. Genealogies now finally seem to be scientifically quantifiable and demonstrable; pedigrees now seem incontestable. In legal and scientific terms, DNA testing has become the means by which to ascertain the truth of a person's identity, supposedly precluding all doubt. On the other hand, however, such genetic techniques have also introduced new insecurities and anxieties. If life can now be artificially manufactured, wherein lies its authenticity? If genetic makeup is extractable, how can it be an expression of individuality? If identity can be translated into information, could this not be circulated, reproduced, and even borrowed or imitated in ways that trouble the fantasies of unique individuality, which, as Katherine Hayles and many others have argued, lie at the heart of the liberal humanism of Western cultures? As Hayles cautions, we already live in a posthuman world in which "information has lost its body": "A defining characteristic of the present cultural moment is the belief that information can circulate unchanged among different material substrates" (Hayles 1999: 1). If Hayles is correct about the dangers of the posthuman age with respect to cybernetics and informatics, what is the significance of her claim for genetic engineering and cloning? What is genetic embodiment, and what new forms of literacy are required in order for such a body to be rendered legible?

Anxieties about the endurance of individuality and authenticity are not entirely surprising in a culture in which Dolly the sheep has already confirmed the possibility of successful cloning, and in which preimplantation diagnosis seems to promise to eradicate the possibility of genetically inherited disease. If the very technology that seemed to guarantee our individuality at the genetic level also promises the extraordinary possibility of our own self-duplication, then the cultural imaginary might well struggle to defend itself against the worst excesses of such a paradox. The present volume specifically addresses itself to the new modes and codes of contemporary technologized forms of life. Throughout this book, a central concern is the question of the reconstitution (or indeed erosion) of the body and its materiality within the cultures of new information technologies and new forms of genetic engineering. This chapter, drawing on theoretical and conceptual frameworks from feminist cultural studies, examines the anxieties surrounding the reconfiguration of the boundaries around the human body, the transferability of its informational components, and the imitative potentialities of genetic engineering and cloning. My intention is to interrogate how the "new genetic imaginary" (Franklin 2000: 197) configures a set of very tangible anxieties about the technological threat to human authenticity and individuality, about the new modes of technological legibility of identity, and about the new codes for deception in the age of replication. I define the "genetic imaginary" as the mise-en-scène of these anxieties, as a fantasy landscape inhabited by artificial bodies that disturb the conventional teleologies of gender, heterosexuality, and reproduction. In the genetic imaginary, posthuman life forms are invented whose histories can be controlled and whose futures might be extended, but who threaten to exceed the controlling gaze of scientific technologies and thus continuously trouble their authority. While many of these anxieties about reproduction, heterosexuality, parenthood, genealogy, and kinship are arguably not new, in the genetic imaginary these fears are amplified through a particular preoccupation with modes of imitation, disguise, and copying.

I begin my exploration of this genetic imaginary by considering two structuring dimensions of genetics that have been widely commented upon by cultural critics. The first is that the gene has no visual signifier.[2] There is no object to be photographed, to be fixed and scrutinized: the gene is invisible, or rather "nonvisualizable." Unlike the fetus or the cell, whose visual image has arguably taken on iconic status in contemporary culture (Franklin, Lury, and Stacey 2000: 19–44), the gene continues to accrue a rather mysterious aura through its enigmatic nonappearance as a visual image. Numerous substitutions, such as the double-helix spiral, have produced metonymic visualizations, but the gene is not a discrete entity and cannot be captured by the visualizing technologies of science and medicine.

The second, and possibly related, structuring dimension of the genetic imaginary is the proliferation of linguistic metaphors as the key trope for how the gene works. Given the absence of a figural image and the power of the linguistic metaphor, we might ask: What forms have constituted the presence of the gene in the cultural imaginary? I seek to answer this question by bringing feminist cultural studies of genetic engineering and cloning into dialogue with my own readings of the current preoccupation with genetics in Hollywood cinema.[3] Of interest here are the cultural forms through which such anxieties have been given a visual life and, in particular, a cinematic life.

TROPES AND FIGURES

In the absence of a discrete object, genetic discourse has proliferated multiple metaphors through which the gene has been imagined, theorized, and brought into public understanding. In Keller's (1995) elegant account of the "metaphors of twentieth-century biology," *Refiguring Life*, the author tracks the shifts in how the gene was metaphorized during the twentieth century, a process that includes the recent radical transformation of biology by computer science:

> It is not only that that we now have different ways of talking of the body (for example, as a computer, an information-processing network, or a multiple input–multiple output transducer) but that because of the advent of the modern computer (and other new technologies), we now have dramatically new ways of experiencing and interacting with that body. . . . It has already been constitutively transformed [Keller 1995: xvii–xviii].

For Keller, practices, knowledge, and metaphors operate in and through one another in such a way that the language of biology not only is embedded in culture but also transforms it in the most material ways. Typically, metaphors have cast genetics as a language that holds the secret of our bodily information. In her study of popular images of genetics, van Dijck (1998: 119) argues that the Human Genome Project is much more than a concerted effort to produce an inventory of the human genome and has entailed "the development, distribution and implementation of a way of thinking about human life." Throughout the 1960s, genes were increasingly referred to in terms that were often interchangeable (as an alphabet, a language, a code, a message, and an inscription), a series of metaphors "rendered more poignant in the popular imagination in the context of the double stranded helix" (ibid.: 36). This model of a genetic code assumed a universal legibility, as van Dijck explains: "If applied to the structure of DNA, 'code' infers the idea of a rule-governed system of communication that can be understood by everyone who has a key to its formative principles" (ibid.). In the 1990s, by contrast with the 1960s, representations of genes moved into a set of associations that combined molecular biology with computer science. For van Dijck, "the gene metamorphosed into the 'genome,' genetics into 'genomics' . . . [and] more intricate mental concepts were needed to imagine the genome as a digital inscription of the body's genetic makeup" (ibid.: 120). This shift meant that genes, rather than being seen as a unidirectional flow of messages or codes, were increasingly conceptualized within more complex and interactional models: "Circulation, rather than a linear flow of information, provided the vector for the dissemination of meaning . . . the body became part of an informational network" (ibid.: 121). This shift in how the gene was conceptualized has had wider implications for how the body in general is imagined. Van Dijck (ibid.: 123–24) summarizes these implications as follows:

> The idea that the human body can be coded in a decipherable sequence of four letters, and hence in a finite collection of information, is based on the epistemological view that computer language—like molecular "language"—is an unambiguous representation of physical reality. Whereas the metaphor of mapping suggests an analog representation—a linear registration of a flat surface—the sequencing of genes ushers us definitely into the digital era. Digital encoding differs from analog recording in imposing a language of zeroes and ones, combined into great complexities, onto the human material body. . . . Through the inscription of "DNA-language" in digital data, the body is turned into a sequence of bits and bytes whose function is no longer exclusively representational.

If, as van Dijck argues, "biotechnology combined with informatics transforms as an (organic) object of knowledge into an ordered collection of biotic components—itself an 'image' or 'concept'" (ibid.: 124), then the problem of reading the body presents itself in a new way. If the gene is not "visualizable," how is it legible? And if genetics is understood as an informational network, what new forms of insecurity enter the dialogic exchange of genetic legibility?

In this shift of emphasis from imagining genetics as a language to imagining it as a code, the geneticized body in contemporary culture moves into what Hayles (2005) has called a "computational universe." This universe operates according to principles not of speech or writing but of code, in which questions of interpretation take on a different significance from those at play in the signification of speech or in the iterability of the written text. If genetic makeup is imagined as a code, how are the possibilities of its transformation, not to mention its transmutation, configured? If the body is a sequence of bits and bytes, how transferable are they? Can we borrow one another's genetic information? How can the informational body's authenticity be anchored, authorized, and guaranteed? How we can recognize authenticity and individuality in what Schwarz (1996) has called "the culture of the copy"? All these questions point to the problem of the legibility of the geneticized body, the technologies of legibility, and the location of the expertise of legibility. With the shift from language to code, how is the concept of legibility itself transformed in the computational, rather than linguistic, universe of genomics (Roof 2007)?

This concern with legibility and duplication brings together debates about digitization and geneticization: in both cases, interference with previously immutable material is at stake. New-media technologies destabilize the authenticity of the image (the photograph no longer necessarily invokes history); new reproductive and genetic technologies potentialize the remaking of the meaning of "life itself" (heterosexual reproduction can no longer claim its sacred hold on nature) (Franklin 2000: 195). But while digitization brings with it endless possibilities for reconfiguring the image, geneticization is a technology without (beyond) an image.

This shift in how the gene is conceptualized has wider implications for how the geneticized body is given visual life. Science fiction cinema has a long history of preoccupation with the presumed predictions of scientific knowledge and with the truths that its visualizing technologies appear to guarantee.[4] In contemporary science fiction film, enactment of the threat of genetic engineering through sartorial disguise has become a common trope. In the rest of this chapter, in order to analyze how the computational legibility of the geneticized body is imagined—or, rather, how fear of the potential problem of genomic embodiment's illegibility is given cinematic life—I shall explore two films that play with such confusions.

CLONING AND CLOTHING

Gattaca (1997) places the threat of visual deception and the problem of the legibility of identity at the heart of its exploration of genetic engineering.[5] Set in "the not too distant future," this science fiction film presents a fantasy of defying the predictions of a genetically determinist dystopia through a story of sibling rivalry in which clothing and cloning have a central but ambiguous relation.[6] The protagonist, Vincent (Ethan Hawke), is motivated by competition with his genetically selected younger brother, Anton (Loren Dean), who has superseded him in his father's favor. Vincent uses an ingenious disguise to pass as genetically superior in order to fulfill his ambitions to become a space engineer and, eventually, an astronaut, at the space station Gattaca. He is, in his own words, a "degenerate" using a "borrowed ladder" to overcome the social exclusion occasioned by his genetic predisposition, in a world where "valids" (those selected from genetically superior embryos) occupy the high-status positions and are valued for their exceptional intellectual and physical attributes, whereas those who have not been preselected, the "invalids" (accented on the second syllable, but a term that also carries obvious connotations of physical inferiority), constitute a low-status labor force performing menial tasks. The substitutability of clothing for cloning is crucially significant here, for this is a story of the impersonation of genetic perfection through a disguise involving both the borrowing of genetic makeup and the transformation of outward appearance. *Gattaca*, an interrogation of what it means to perform identity in an age of genetic engineering and genetic therapy, explores the fissures between and among physical appearance, how physical appearance is read by others, and the supposedly unique genetic coding that constitutes individuality. The invalid Vincent, deploying expertise in sartorial disguise, physical alteration, and prosthetic DNA, successfully impersonates the more genetically desirable "valid" Jerome (Jude Law), who, ever since an accident, is no longer the top athlete he once was. Jerome's genetic perfection is sold as a commodified identity to Vincent, who develops elaborate rituals of shedding his own genetic traces (skin, hair, nails) and adopting Jerome's (blood, hair, urine) on a daily basis. The combined labor and ingenuity of Vincent and Jerome produces an impostor clone of genetic perfection, who eventually outwits the supposedly superior "valid" Anton. Here, sibling rivalry motivates the use of impersonation as a way to prove the limits of genetically determined social injustice.

The film thus explores a fantasy parallel to that of the child in the case related by Phillips (1998): How could genetic duplication overcome filial displacement? And how might the artificial production of similarity guarantee the fulfillment of desire? And yet the narrative suspense revolves around the risk, rather than the safety, of successfully achieving similarity. Here, disguise is mobilized to challenge genetic determinism, and the threat of discovery is placed at the heart of the narrative structure. Every time Vincent enters the space station, his genetic identity is screened; thus, in order to imitate Jerome's genetic perfection, Vincent has to become both master of the image and master of deception. His brother's contrasting role, as a detective who uses genetic screening to fight crime, contributes to the narrative tension, particularly when a murder at the space station Gattaca requires him to screen all its employees, including his brother. Vincent's disguise as Jerome presents a cinematic vision of cloning, in which the technologies of imitation are set in battle against the technologies of genetic engineering and genetic screening. Through its exploration of the dynamics of genetic disguise, *Gattaca* poses a series of complex questions about sameness and difference and about the authenticity of identity in an age of replication. It poses the question of how genetic engineering "troubles" identity, and vice versa.

The eugenic fantasy of *Gattaca* is conveyed through a mise-en-scène that emphasizes repeating patterns and sequences. The opening sequence appears as a highly formal display that mimics the genetic sequence from which the title of the film is derived. (The title is composed of a combination of the first letters of each of the four bases of DNA—adenine, guanine, thymine, and cytosine.) The titles also highlight the same four letters as they appear in the individual names of the actors, and the marked letters remain onscreen for a few seconds after the name's other letters have disappeared, leaving a lingering visual imprint (thus the name of Ethan Hawke momentarily becomes *ET A A*). The mise-en-scène throughout the film, like a visualization of genetic information, also produces an aesthetic premised on symmetry, order, and control. As Kirby (2000: 204) has written, the film "visually . . . conveys an antiseptic world that has been purged of imperfections. . . . The Academy Award–nominated sets, whether at the Gattaca Corporation or in Jerome's . . . apartment, show a sterile and blemish-free world filled with smooth stainless-steel surfaces." According to Romney (1998: 48–49), the film's visual style "pushes its cultivation of surface elegance to parodic excess," but ultimately it produces the visual equivalent of the genetic technology it seeks to caution us against: "The film [tries] to have it both ways—tut-tutting at *Gattaca*'s body fascism while asking us to feel for leads as pedigree-bred as Uma Thurman and Ethan Hawke. . . . You begin to suspect that production decisions were made to strictly eugenic criteria."

The film throws into particularly sharp relief the question of the status of visual evidence in the age of genetic engineering and cloning. In the absence of a figurative image of the gene, magnified shots of the human body stand in for genetic information, as if being visually closer to the body might offer a depth of knowledge about its new genetic truths. Appearing to render DNA visible, close-up shots of familiar bodily surfaces and fluids repeatedly symbolize corporeal legibility. Familiar scientific techniques through which identities are habitually translated into visual evidence (the fingerprint, the blood test, the infusion) are shown in close-up as if to give genetic information a visual presence. Throughout *Gattaca*, close-ups of fragments of the human body refer back to the title sequence, where microscopic modes of spectatorship produce a sense of striking proximity to the image. But they do so ambiguously. Seeing the body (Vincent's) so close up, and yet misrecognizing it so profoundly (as Jerome), invites us to contemplate the relationship between seeing and knowing, between observable corporeal surfaces and the identities beneath them, for if close-up shots offer the promise of truth through a magnified image, ultimately they do so here only to undermine a sense of certainty. In this film, more vision often leads to less knowledge—to distortion, misreading, and even reversals of perception. And, as the whole trajectory of the film demonstrates, visual information cannot always be read as transparent proof of genetic identity, for both visual evidence and genetic evidence are open to manipulation and are susceptible to the indeterminacies of interpretation.

If a masculine desire for mastery of the technologies of detection and deception lies at the heart of *Gattaca*'s challenge to genetic engineering, *Species* (1995) explores the terrifying failure of scientific technologies to control the drives of a genetically engineered female who combines the duplicity of the traditional femme fatale with the devouring appetite of the "monstrous feminine" (Creed 1993).[7] As the advertisement for the film puts it, "Men cannot resist her. Mankind may not survive her." In contrast to *Gattaca*, in which masculine perfection is technologically engineered with an almost military precision, reaping the rewards of professional ambition and defying the potential constraints of an unethical eugenicist science, *Species* exposes the fallibility of scientific (and military) technologies in their quest to contain the excesses of a "posthuman" female reproductive drive whose only ambition is to find a genetically viable mate and thus guarantee the future of her species. In both films, however, the artificiality of genetic engineering is tied to an exploration of engendering as disguise: gender identity is perfected, worked on, borrowed, copied, and imitated. In *Gattaca*, masculine genetic perfection (Jerome's) is purchased, manipulated, and impersonated by an inferior other (Vincent); in *Species*, feminine sexual desirability is copied, stolen, and performed with computational accuracy by a genetic mutant on an urgent reproductive quest. In the first film, masculine perfection as a genetic identity can be approximated by a "degenerate" human through imitative technologies; in the second, feminine perfection is a distributed cultural code that is reprocessed as a database resource by a transgenic species for her own disguise as fully human. In both films, the interplay of the legibility of codes—biological, informational, and cultural—is crucial.

Species presents the story of a genetically engineered female called Sil who escapes from a U.S. base (in Utah) where scientists have created her by artificially combining human DNA with DNA mysteriously received from a supposedly friendly source in outer space. The alarmingly rapid development of this apparently human female—the embryo is formed in hours; the fetus, in days; the four-month-old child resembles a six-year-old—is observed by scientists through the transparent glass cell in which Sil is kept for security. The scientists, having witnessed Sil's physical responses (shuddering and rapid eye movement) to what they assume to be nightmares (but which later turn out to be more like interspecies memories), decide to terminate her by poisoning her with cyanide. With its echo of Holocaust atrocities, this gassing of an innocent child by a ruthless regime run with military precision is observed by the head scientist, Xavier Fitch (Ben Kingsley), whose panoptic vision of the child's suffering is mediated by a series of screen visions of the scene. At the moment when her life is threatened, the apparently helpless blonde child becomes a bionic force; she escapes by diving through the glass walls of her cell, and she runs off the base to freedom. The rest of the film shows her rapid metamorphosis into a posthuman-species adult (Natasha Henstridge) whose violent quest to seduce a "genetically desirable" man and reproduce biologically is depicted as a battle against time, as a task to be accomplished before the team of experts assembled to capture her can prevent the reproduction of this alien species.

The team members represent the combined strength of their human qualities (physical, technological, intellectual, and emotional), in contrast to Sil's posthuman drives and capacities. Each member of the team, like a representative of the diversity of the human gene pool,[8] is defined differently across the axes of gender, race and class. Preston Lennox (Michael Madsen) is a white working-class tough guy. Laura Baker (Marg Helgenberger) is a white middle-class scientist. Steve Arden (Alfred Molina) is a naive white middle-class intellectual (an anthropologist). Dan Smithson (Forest Whitaker) is a sensitive middle-class African American empath. Each member of the human team connects in some way with Sil: Lennox shares her brutality and her ingenuity; Laura is twinned with Sil's white, blonde femininity and shares her desire to find a male partner; Steve is ultimately her genetic match and successful reproductive victim (the father of her offspring); Dan's compassionate emotionality contrasts directly with its complete absence from Sil's responses to her victims (one of multiple markers of her posthuman status).

The mise-en-scène throughout the film is organized around a preoccupation with screens, vision, and the problem of transparency. Following the generic conventions of science fiction film, the power of the scientists' visualizing technologies in *Species* is tested against the ingenuity of the threat to humanity, a threat that science itself has produced and must now destroy. The high-tech laboratory of the film's opening scene displays multiple sources of visual control and screening (the transparent cell, computer monitors, the panoptic architecture of observational regulation) as well as the spectacular disruption of this relay of screen controls by a young girl turned bionic escapee. The narrative tension that follows is built around the battle between, on the one hand, the combined expertise of the team hired to identify and capture Sil (using military tracking devices and digital visualizing techniques alongside human instinct, intuition, and brute force) and, on the other, Sil's rapidly expanding databank of resources for disguise and deception (using images from popular culture and the sartorial styles of other women alongside her bionic physical power, her transmutational body, and her alien instinct for making a genetic match to maximize the preservation of her species). The power of science to make visible that which it seeks to experiment on and to control is pitted against Sil's instinctive power to disguise that which she seeks to conceal from the scientists. Thus the dangers of genetic engineering are once again articulated through a mise-en-scène of screening devices (visual, informational, and instinctual) through which each side mobilizes its "reading skills" to make the other more transparent. The artificiality of Sil's genetically engineered design is paralleled by the artificiality of the image cultures she encounters through multiple screens: that of a stolen portable computer on a train, the shop windows along the streets of Los Angeles, and the television screen that is the site of her channel surfing in her motel bedroom.[9] These visual images, as random as the stolen food on which Sil binges before her metamorphosis, bombard her (and us) with a dizzying array of arbitrary imperatives about how to "do" femininity (her appetite for both—the images and the food—is driven equally by a biological imperative).[10] This parallel is underscored by the simultaneity of Sil's accelerated biological mutations (from girl to adult woman during a single journey by train, through a monstrous, chrysalis-like metamorphosis) with the multiple cultural mutations (improvised feminine desirability through commodity consumption) that facilitate her disguise.[11] Again, cloning and clothing coincide. Sil becomes a mistress of the masquerade, stealing and borrowing whatever cultural accoutrements she needs (money, credit cards, cars, identities, clothes) as she appears as a train conductor, a bride, a sexual predator. For Sil, femininity is strategic, and its visual imaging in popular culture is simply a database for random reassemblage.

But Sil's "computational universe," in which she reductively reads everything as data, is also her downfall, exposing her outsider status and leaving behind clues for the team of experts to follow. She reads the artifice of femininity too literally, as if it were solely an informational code; she treats images as pure information, ignoring the contextuality that confirms the dialogic relationality of their forms.

For example, imitating a seductive gesture she has seen on television, Sil removes her top in a public space in front of a man she decides to mate with. Sil's inability to imitate femininity in an appropriate way (she wears a wedding dress to go shopping) exposes the limits of her mode of reading culture as information, for although femininity can be acquired through the repetition of conventions, it still has to become an ontology before it can be correctly performed. Sil reads femininity merely as surface image, not as embedded cultural practice, and so her cultural illiteracy contrasts with the biological "literacy" that she deploys with unfailing intuitive accuracy. In her quest to find a genetically suitable mate, Sil reads the candidates' genetic makeup by simply sensing their DNA through physical proximity, with a kind of instinctual drive for the survival of the fittest (she rejects the imperfect ones before reproductive intercourse). Sil, as the cultural ideal of traditional feminine beauty (she is white, blonde, slim, tall, and young), is guaranteed a heterosexual mate. Utterly strategic, she performs, with an almost masculine detachment, the femininities necessary to her reproductive quest. Here, conventional sexual desire is replaced by a reproductive drive that is motivated by a satisfactory genetic reading of biological data: the body is simply a code to be read. Sil misreads the images of femininity screened in consumer culture, seeing them as nothing but informational codes, but she reads male bodies as genetically transparent.

For the team of experts, the contrast between the legibility of genetic embodiment and the illegibility of cultural codes is played out in the opposite direction. The team members, motivated by their desire to see beyond the mask of the genetic combination of human/species—to see the true beast within Sil—decide to "grow this creature with just its own DNA . . . to investigate its vulnerability . . . to see it without the camouflage," as Laura says to her colleagues. In the scene where this is attempted, there is an abrupt cut from close-ups of the pornography and the advertisement for hair coloring that Sil is watching on television to a close-up of a computer-screen image of a microinjection, placing images of biological and cultural artifice into deliberate dialogue. The desire to master life, articulated through a mise-en-scène of technoscientific apparatus at its most digitally powerful, is enacted simultaneously on the computer screen and on the cinema screen. The view of the microinjection—which is being conducted by remote control on a cell containing alien DNA that is in a sealed unit with a transparent wall—is magnified on one screen through its projection onto the other. The male members of the team are led by the only female among them, the molecular biologist Laura, and the team as a whole is shot from above, the members turning their heads one by one toward the computer screen to witness what they, and the cinema audience, expect to see with their own eyes: genomics in action. As Sarah Franklin (2002) argues, the now familiar close-up of the prosthetic penetration of a human cell by a pipette (accomplished via the directing hands of the scientist) has become an iconic emblem of the new possibilities of genetic engineering.

But as the female scientist narrates the story of the new life form she is engineering, the camera fails, and both the computer screen and the cinema screen go blank. The diegetic and nondiegetic spectators share a loss of point of view in the moment of technological failure. When the screen goes blank, we assume that the procedure has also failed. In other words, we equate the failure of visual technology with the failure of genetic engineering. During the time it takes to replace the camera in the sealed unit containing the alien DNA, neither audience can see the state of the microinjection procedure. When the computer's camera is switched back on, the two audiences share the simultaneous shock effect of the successful penetration of the cell by the alien DNA and its monstrous outcome—the rapidly dividing alien cells, which resemble an amorphous brown cellular mass. Accompanying the visual horror of this frothing cancerous growth is an equally powerful sense of disavowal that the procedure could have been completed without our knowledge, and beyond our vision.

Genetic engineering in *Species* is a form of biological design whose artificiality is presented as analogous to the cultural disguise of impersonation, and which produces similar problems of illegibility. Just as Sil fails to impersonate human femininity in a fully convincing manner, the scientists in the film are thwarted in their attempts to imitate biology, to control life through genetic engineering. Sil's literal readings betray the truth of her genetic artificiality; the experts' inability to see the moment of the microinjection of the alien DNA betrays the fallibility of the humans as well as of their technology: despite all the available visualizing technology, they cannot see what is really going on at the genetic level. In the moment of technological failure, their (and perhaps our) desire lingers, exposed. The relay of human and alien misreadings throughout the film shows each side's respective failure to shift appropriately among the different registers of legibility required by the codes and modes of culture, biology, and technoscience. The scientists misread transparency as certainty (the transparency of Sil's glass cell does not guarantee security), revealing their misplaced belief that knowledge and vision guarantee each other. Sil misreads culture as pure information (images, bodies, and clothes are all data resources), and her misreading suggests that her genetic makeup translates into a computational subjectivity which perceives the world as bytes of life.

SCREENING AND SCREENING

The story of linguistic and conceptual substitution (of clothing for cloning) with which I began this chapter enacts the problem that lies at the heart of the genetic imaginary: the desire for transparency produces a proliferation of confusion. In displacing sameness at the genetic level onto sameness at the sartorial level, the child reads biological depth as cultural surface. Duplication of genetic makeup (cloning) is imagined through duplication of dress (uniform) in such a way as to make transparent the absence of difference that might mark out some individuals (like the child's sister) as special. Sameness spells safety if it can be made visible. If kept in sight, the transparency of sameness can be guaranteed. In this fantasy of "doing clothing" when she grows up, the child reveals her wish to control difference (making people wear the same clothes, like her headmistress does) and to have "everything" (being in two places at once) by cloning herself. The wish to abolish the threat of difference enables her to imagine her own desire fulfilled. If heterosexual reproduction can be bypassed ("you don't need a mummy and a daddy") in favor of self-replication through the new paternal authority of science ("you just need a scientist," a "man") in a place where individuality is no longer a threat ("no one's special"), then the child can fulfill her desire to have "everything": she can be in charge of her own destiny and yet blend into the crowd. While her individual agency is guaranteed through reproductive authorship (like a male scientist, she will make all these babies), the threatening individuality of others (such as her sister's capacity to entrance their mother) is disavowed through the new potentialities of replication in genetic engineering. This imagined shift moves her from powerless child (subject to the authority of school and family) to controlling adult (she will regulate reproduction, uniforms, and parental affection). In her fantasy, sartorial uniformity, like its genetic counterpart, will provide the justice of transparency through new and visible laws of sameness.

But this young girl is not alone in her confusion. Her story may reveal her own particular unconscious investments in overriding traditional family structures, but the substitution of clothing for cloning has wider imaginary purchase, as this chapter has shown, for the substitution stages the displacement of the biological onto the cultural so as to make transparent the hidden threat of sameness in the techniques of the new genetics. Like the child with her fantasy of cloning, the genetically engineered girl in *Species* has no biological parents but is instead originally created by a male scientist (Fitch) who narrates the story of her miraculous conception. And, like the conflation in the child's fantasy, Sil's genetically engineered duplication is expressed through her clothes: her transgenic mutations are echoed through her endless changes of identity, carried out with sartorial disguise. Sil's ideal female body is the register for confusions around legibility; the disguise at her disposal enables her to camouflage the "monstrous feminine" through the duplicity of the femme fatale. What we might call her mutational femininity articulates her genetically engineered embodied form.

In *Species* and *Gattaca* alike, the image of the body is the site of deception, echoing the dangers of interference at the deeper genetic level. The moments of apparent visual certainty on the screen (the microinjection of alien DNA to decamouflage Sil's disguise, and the appearance of Jerome's identity card through the blood test of his DNA provided for the police by Vincent) rehearse what we might call the scientists' desire for genetic transparency: the power to visualize pure information. The impersonation of masculine perfection in *Gattaca* relies on cultural disguise to literalize the problems of biological manipulation. When Vincent adopts Jerome's identity, he produces himself as an image, and as a genetic data set to be read (or misread) by screening technologies. In both films, however, transparency eludes the screen.

Genetic engineering is imagined to pose a threat to the transparency of identity—to its legibility by experts, by sexual partners, by colleagues. In different ways, each film investigates the potentialities and limits of visualizing technologies and information technologies to make legible the genetic truths of identity. The power of human ingenuity and the tools of cultural design are pitted against the power of genetic engineering and biological design to transform the fundamental meaning of life itself. In *Gattaca*, a deceptive double's skills of impersonation prove that the attempt to extend surveillance to the realm of genetic scrutiny is an impossible and undesirable project. In *Species*, a monstrous femme fatale develops techniques of cultural disguise paralleling her own biological conception, techniques that put her one step ahead of the experts who designed her. The potential for deception is placed at the heart of both narratives, which juxtapose the scientists' power to guarantee authenticity with the endless potential for interference, imitation, and duplication.

If the confusion of cloning with clothing symbolizes the desire to make visible, on the surface of the body, interference at the hidden, genetic level, the status of the image is crucial to the structuring anxiety that is at stake (and to the reassurance that is sought). The cinematic language of the gene searches for substitutions (in the absence of visual signifiers) while simultaneously cautioning against the certainty that the image promises. Both *Gattaca* and *Species* question the power of visualizing technologies to secure genetic certainty. But in neither film can genetic information be made transparent through screening techniques (in all senses of the term). In the requirement that others complete the dialogic circuit of identity production, the troubling elements of competing interpretations and varying registers of legibility become apparent. In both films, the desire to transform the body into legible information that can be read as pure data and replicated accordingly—the desire to contain the threat of sameness by making it visible—is thwarted by unexpected uses of technologies of imitation. In such a genetic imaginary, anxieties about the legibility of corporeal surfaces and depths are played out across a matrix of informational and biological registers. Meanwhile, the transparency of the screened body remains a dubious and unfulfilled wish.

NOTES

1. For a lucid account of dialogic theory, see Pearce (1994).

2. For important analyses of cultural constructions of the gene, see Nelkin and Lindee (1995), Turney (1998), van Dijck (1998), and Franklin (2000).

3. Genetic engineering has long since been part of Hollywood cinema's preoccupation with artificial bodies; examples include *The Fly* (1958), *The Boys from Brazil* (1978), and *Blade Runner* (1982). More recently, however, there has been a proliferation of films about new forms of genetic engineering, including cloning; examples include *Jurassic Park* (1993), *Species* (1995), *D.N.A.* (1997), *Gattaca* (1997), *Alien Resurrection* (1997), *Species 2* (1998), *The Replicant* (2001), *Code 46* (2003) *Godsend* (2004), and *The Island* (2005). For a fuller list, see Nottingham (2000).

4. Critical work on the genre of science fiction film includes Kuhn (1990, 1999), Penley, Lyon, Spigel, and Bergstrom (1991), Sobchack (1997), Telotte (1999, 2001), Newman (2002), and Wood (2002).

5. For a more extensive reading of *Gattaca* in relation to feminist and queer theory, see Stacey (2005).

6. For important discussions of *Gattaca* in the context of the history of science fiction film, see Kirby (2000) and Wood (2002: 167–75).

7. There are now two volumes of feminist work on the femme fatale in film noir; see Kaplan (1980, 1998).

8. I am grateful to Maureen McNeil for raising this point in discussions about the film.

9. For an excellent analysis of the multiplicity of screens in postmodern culture, see Friedberg (1993).

10. An important analysis of the current "teratological" fascination with mutational bodies in Hollywood cinema is offered by Braidotti (2002).

11. Sil's spectacular "monstrous" mutational body was designed by H. R. Giger, who also designed the special effects for *Alien* (1979) and *Alien Resurrection* (1997).

PART 3

■ *Remediated Bodies*

8 ▪ MyLifeBits

The Computer as Memory Machine

JOSÉ VAN DIJCK

An old friend recently admitted, with a sense of understatement, that the size of his personal digital collection had outpaced his ability to keep track of its contents. Since acquiring a digital photo camera and a scanner in 1995, he had taken, stored, and scanned well over a hundred thousand pictures of his daily life, work, and family. His collection of DVDs and audio files also confronted the fate of infinite expansion, due to the increasing availability of peer-to-peer technologies. The act of recording and storing files, images, audio, and data, combined with occasional camcorder activity and heavy Internet use, absorbed almost every minute of his spare time. Space was no longer an actual or virtual constraint, since only my friend's film DVDs were still stored as material artifacts on the shelves, and computer RAM has become a bargain. His proposal to transfer the family's entire collection of old photographs and videotapes onto digital media had met some resistance from his partner, who expressed her attachment to the touch and feel of analog products. Digitization had also confronted him with issues of time and order: When would he have time to enjoy and relive all these recorded cultural and personal moments if he was constantly engaged in capturing and storing the latest experience? And what order would allow him to retrieve specific moments, since the danger of their getting lost in his multimedia repository of personal memories was growing by the day?

Digitization is surreptitiously shaping our acts of cultural memory—the way we record, save, and retrieve our remembrances of life past.[1] With the emergence of every new technology, from print to photography, and from the gramophone to the computer, people have hailed and despaired of new means of (self-)recording, storage, and retrieval. Since the 1960s, the shoebox with its variety of private documents (photos, letters, diaries, home videos, voice recordings, and so on) has expanded to a giant suitcase, or an attic. In addition, personal collections of recorded cultural content (audiocassettes, videotaped films, taped television programs) are cherished as a formative part of autobiographical and cultural identity; they typically reflect the shaping of an individual within a historical time frame. Together, private documents and personal collections of cultural content constitute what I call "mediated memories": memories recorded by and (re)collected through media technologies.[2] These technologies are never simply machines, since they are always firmly embedded in the contexts of cultural practice and defined by the cultural forms they engender. Few scholars have bothered to theorize the power of media technologies to shape the materialization of cultural memory, a shaping power that is particularly discernible in periods of technological transformation (Hoskins 2001; Gross 2000).

The gradual takeover of analog by digital technologies marks a new chapter in the mediation of cultural memory.[3] Digital technologies offer new opportunities but also introduce new complexities into people's everyday lives; the computer, with its ever-expanding capacity for memory, is rapidly becoming a giant storage-and-processing facility for recording and retrieving "bits of life." As more people discover the pleasures of digital recordings and presentations, many, like my friend, will also come to acknowledge the problems that accompany new technology, such as the issues involved in handling exploding quantities of personal data. The remediation of old and familiar forms or genres, such as family albums, handwritten letters, and scratched compilation tapes of favorite tunes, foregrounds the significance of materiality in the process of remembering. More implicitly, it raises questions about the various gendered roles in the collection and storage of a family's heritage. To replace common analog forms, commercial enterprises are quick to offer such alternatives as family albums of digital photographs, or formatted weblogs (digital diaries, known as blogs) or scrapbooks, to facilitate storage and retrieval. Not surprisingly, these enterprises tend to focus on the products of memory, turning mediated expression into prefabricated exercises that are based on conventional analog genres. Software engineers and companies have recently started to address the question of storage by designing digital tools to accommodate the infinite expansion of our digital memories. Some projects simply promise to solve the urgent "shoebox problem"; others also purport to design completely new systems of memory storage and retrieval; yet others boast that their new software and hardware will revolutionize our very ability to remember.

Although we tend to attribute not just the bliss but also the dilemmas of expanding collections of memories to digitization, the dilemmas are neither completely new nor uniquely related to the computer age. A more interesting question, in my view, is how digital technologies, by changing the material basis of our mediated memories, (re)shape the nature of our recollections and the process of remembering. Simply put, how do digital technologies affect acts of cultural memory? How do they frame new ways of collecting, storing, and retrieving mediated records of the past? How do they reshape the social, often gendered, use of memory tools? In this chapter, I argue that digital technologies, rather than revamping the products of memory, tend to change the performative nature of memory—that is, the way we create and deploy memories as a way of giving meaning to our lives. Digital technologies may introduce tentativeness as a stage in the memory process. They may prompt a multimodal sense of remembering. They may reconnect memories of the self to the reflections of others or to reports of events in the world at large. Finally, they may alter gender-specific roles in the creation of a personal heritage.

FANTASIES OF A UNIVERSAL MEMORY MACHINE

Fantasies of the perfect memory machine have always accompanied the invention of new media technologies. As McQuire (1998: 127–38) points out, the splitting off of living memory from so-called artificial or technological memory, first made possible by the invention of writing, engendered dreams of complete recordings as well as of systematic ordering and retrieval of lived experience. The German philosopher and mathematician Gottfried Wilhelm Leibniz (1646–1716) and the English mathematician Charles Babbage (1792–1871) are both credited with having envisioned mechanized memory tools, and these have been seen, in hindsight, as early precursors of the computer. Yet the most famous visionary of the modern memory machine was undoubtedly Vannevar Bush, former director of the American Office of Scientific Research and Development, whose fantasies had more than a slight impact on the ideas of contemporary engineers and scientific communities.

In his famous 1945 essay "As We May Think," Bush expressed his fear that society would soon be bogged down by explosive growth in specialized publications, and he urged scientists and engineers to turn to the massive task of making our bewildering store of knowledge more accessible. Placing himself in the tradition of Leibniz and Babbage, who envisioned extensions of the mind in the form of calculating and arithmetical machines, Bush committed himself to designing a memory machine that would enable the storage and retrieval of various types of records: documents, photographs, films, television programs, and recordings of music and speech. He predicted the emergence of a new type of machine, one that would allow humankind to avoid repetitive memory tasks, and that would thus make room for more creative thought. Analogous to the idea of a calculating machine was the "memex," which Bush (1945: 14) defined as "a device in which an individual stores all his books, records, [and] communications, and which is mechanized so that it may be consulted with exceeding speed and flexibility." An essential feature of the memex was its ability to automatically select and retrieve any stored item swiftly and efficiently. Bush's fantasy of the memex has been hailed as the greatest vision to have anticipated the computer age, but it has also been criticized for its ideological undercurrents of pioneerism and frontierism (Kitzmann 2001). For the purposes of my argument here, I am less concerned with Bush's general ideas about science and progress than with his assumptions about the relationship between the human brain and the memory machine. Bush proposed to model his memex after the human brain, in order to artificially duplicate the mental process of memory retrieval, thus relieving the brain from a number of repetitive tasks. According to Bush, conventional systems for storage and retrieval, which classify data alphabetically or numerically, and in which one locates information by tracing it through subclasses and indexes, are cumbersome and counterintuitive. The human mind, he wrote, operates by association, and so should memory machines.

Bush's concept of the memory machine is both mechanical and paradoxical. It is mechanical because he presumes an unambiguous vector between technology and the human mind: the memex ought to function as a human mind. Memory, to his regret, is fallible ("transitory"); therefore, a machine should take over part of the brain's function, to prevent amnesia due to information overload. Ideas, memories, and thoughts are stored in documents or other recordings, and these recordings can be randomly retrieved. Writing about data, Bush equates "bits of information" with "ideas, memories, and thoughts" that can be put away in a repository and be retrieved randomly or by association. Nevertheless, the retrieval of documents from a database and the retrieval of memories from a human brain are fundamentally different processes with very distinctive goals. Documents or recordings can be stored in a database, and we want them to be there unchanged as we retrieve them and subject them to (re)interpretation; memories, by contrast, are never unchanging "data" that can be stored and retrieved in "original" shape. Or, as Winkler (2001: 103) puts it, "Material storage devices are supposed to preserve their contents faithfully. Human memories, on the other hand, tend to select, reconfigure, and forget their contents—and we know from theory that this is the real achievement of human memory. Forgetting, in that sense, is not a defect, but an absolute[ly] necessary form of protection."

And Bush's concept of the memory machine is also paradoxical. He states that machines should be modeled after the human brain, but he does not account for the fact that the brain interacts with a machine, and vice versa; therefore, in Bush's vision, there is no room for the logic that technology structures as well as reflects cognition—a serious flaw echoed in quite a few contemporary hypertext theories. Kendrick (2001: 231) has critically analyzed how notions of hypertext are structured analogously to the mind, thus promoting the "connection between a technology of links and nodes and the presumed associative ability of the mind." Hypertext enthusiasts also proclaim that interactive software programs erase mediation, liberate the writing subject, and empower writers who were formerly restricted by the constraints of narrative discourse. Kendrick counters the commonplace notion that hypertexts, being more like the human mind, are more natural (hence better) with the notion that technologies of inscription actually shape human cognition. Hypertext gurus, however, like Bush, regard the human mind as the model for the machine rather than seeing the mind in interaction with the machine.

The presumed division between memory and machine—between lived experience and mediated experience—finds fertile ground in a notion popular among cybergurus: that brain functions can be taken over by digital equipment, thus disembodying the mind. As Smelik points out in chapter 9 of this volume, the masculine fantasy of transgressing physical boundaries to achieve the ultimate heights of virtual reality underlies the plots of quite a few cyberspace movies. Such a desired state of immateriality and disembodiment can also be traced in Bush's notion of the perfect memory machine: detached from the body, the memory machine becomes a masculine model for total recall and, at the same time, for total control. The disembodied memory machine stands in unarticulated opposition to the "embodied" notion of remembering, according to which the use of memory devices is seen as one part of the very personal act of creating, storing, and retrieving memories. Not coincidentally, the activity of making family albums, storing pictures, and updating the family's records is typically a female activity that calls for a personalized, embodied take on our conceptualization of memory and memory machines.

Bush's fantasy of the universal memory machine, the memex, has inspired many recent projects concerning the inscription, storage and retrieval of personal memories or "bits of life." In its 1945 utopian form, Bush's concept prefigured the need for an exterior digitized memory with unlimited capacity. It anticipated the transformation of personal collections into multimedia compilations of images, text, and sound—technical tools for evoking memories. And, most of all, the memex foreshadowed the need for automated retrieval systems as a consequence of exploding quantities of information. In sum, the memex fantasy contained everything needed to solve the "giant shoebox" problem, and so it is hardly surprising to find that many recent projects have been inspired by Bush's fantasy. More important, though, is the fact that his mechanical and paradoxical assumptions about the machine's resemblance to the brain are echoed in contemporary digital projects, although the terms of this proposition have been reversed. In the pages that follow, I discuss a specific contemporary project for designing a digital memory machine, a project that illuminates how the brain is no longer envisioned as the functional model for the computer, and how the computer now serves as a standard for the workings of human memory.

OF DIGITAL SHOEBOXES AND JUKEBOXES

The past decade has witnessed the emergence of various initiatives, commercially as well as privately funded, to address the complex management of digital personal collections. Despite differences among the goals and outcomes of these various projects, their common assumptions, which involve the functioning of personal digital collections in relation to memory, betray a peculiar desire to control and manage the human brain as if it were a computer system. These assumptions also demonstrate how the software design is inimical to any view of computers and the organization of personal memory as being interconnected. An extensive description of the various projects would be beyond the scope of this chapter; instead, I will concentrate on the analysis of one specific project, MyLifeBits.[4]

MyLifeBits was launched in 2002 by the Media Presence Group, led by Gordon Bell, at the Microsoft Bay Area Research Center, San Francisco.[5] The project leaders work on a comprehensive software system while simultaneously communicating their goals and mission to a broad audience.[6] The Microsoft engineers aim to build multimedia tools that allow people to chronicle the events of their lives and make them searchable, since memories deceive us: "Experiences get exaggerated, we muddle the timing of events, and simply forget stuff," says one of the project leaders. "What we want and need is a faithful memory, one that records and builds on the reliability of the PC."[7] In the many interviews that Gordon Bell has given to the news media, he pitches MyLifeBits not only as the solution to the "giant shoebox" problem but also as an organizer of life: everyday events will be fully recorded in text, images, and audio and stored in an orderly fashion on a computer. Each item will be tagged by audio or text annotation, and the tags themselves may also be cross-linked. Since 1999, Bell has been placing all his personal "bits of life," including his parents' photographs, onto a hard drive, to test the program. MyLifeBits, he claims, is more than a memory; it's "an accurate surrogate brain," the realization of Vannevar Bush's memex machine, which, like MyLifeBits, also featured automated retrieval as its highest ambition. In the future, Bell imagines, a compulsive recorder will be able to call up a single day in her life and get an hour-by-hour breakdown of what she did, said, and saw. One major advantage of the MyLifeBits software, in addition to its power of chronological retrieval, is its ability to allow a "Google-like search on your life," Bell says—to enable retrieval of random memories via typing of a verbal tag.

The MyLifeBits software starts from the notion that stories based on artifacts (for example, photos) are tools for personal memory. The "shoebox" of digital items is viewed as a nonhierarchical repository of annotated data from which the user constructs a story every time she or he retrieves a single bit. Annotated stories can be browsed in the same way that material on the World Wide Web is browsed; that is, the user, by means of one or more keywords, simply follows the links connecting one resource to another. Thus memory is conceived as an associative journey through linked "bits of life," which later can be re-presented in any mediated form (for example, as a PowerPoint presentation, a slide show, or a photo album). It is not the stored items themselves, but rather the story presentation enabled by item retrieval, that comprises the conceptual heart of this project. Our children's or grandchildren's most valuable inheritance, Bell claims, is not a shoebox of assorted items but a selection and representation of annotated stories. As one of his interviewers concludes, "Your cinematic deathbed flashback will already be uploaded to your hard drive."[8]

In MyLifeBits, we see the idea of the computer as a model for the brain extended from the computer's storage capacities to its capacities for retrieval and presentation. According to this model, not only does the human mind work like a computer, it ought to work like the World Wide Web, presumably the most efficient and effective navigation environment to have emerged in the digital age. Google, the celebrated Web-based search engine, becomes the prime conceptual model for running and activating the human mind; the Googlization of memory turns a brand name into an everyday practice. It is also interesting to notice how the Microsoft engineers construct the notion of life as a story and simultaneously equate life stories with mediated formats: personal memories cast as narratives, using images as material signposts, conceptually morph into preformatted media presentations—preferably copyrighted by Microsoft, of course. But the understanding of memory in terms of media formats is by no means an invention of the digital age. Ever since the late nineteenth century, we have been using the metaphor of the panoramic flashback to indicate a person's hovering in the twilight zone between life and death;

in so doing, we have actually been projecting a cinematic *procédé*, including montage, slow motion, and black-and-white shots, onto a presumed physical-psychological process (Draaisma, 2002). The metaphor of film organizes its own perspective on the construction of memory, and so does the preformatted "cinematic deathbed flashback" that the MyLifeBits software has already placed on your hard drive. In the funeral business, not surprisingly, it is no longer a novelty to use a PowerPoint slideshow in presenting a review of the deceased's life (Moran, 2002: 89–93). The digital secularization of memorial rituals could become much easier if you anticipated your life's end by annotating its prelude in the course of living.

The design of MyLifeBits deftly reflects (and smartly caters to) two contemporary anxieties—about managing one's life, and about amnesia. For an upscale Western audience, the notion of managing data has become an attractive metaphor for controlling life. To have an experience at a date and time of one's choosing—rather like watching a television program that has been recorded on a VCR—takes some pressure off life's fast pace, regulated by the clock. What could be more appealing to a contemporary user, struggling with time constraints in an experience economy, than the storage of events in the form of mediatized, retrievable memories? Anxiety about missing the experience of seeing one's children grow up can be assuaged by the thought of a personal memory machine that allows precious moments to be replayed at a time more convenient than the ever-demanding present. Thus experiences etched in the dimension of time become a timeless repository of reruns. As for amnesia-related anxiety, it feeds on the prospect of such harrowing memory disorders as Alzheimer's disease. Complete storage of one's personal memory and collections should prevent the erasure of one's unique identity. And if anxiety about forgetting is inextricably intertwined with anxiety about being forgotten, then, as MyLifeBits insists, the major beneficiaries of the software are one's descendants. Immortality through software cultivation seems like a sound investment.

Projects like MyLifeBits capitalize on the digital enhancement of limited human memory. Because they view human memory as profoundly lacking in its prime function—to remember in full, and exactly, registrations of past events—they tend to focus on those products of memory that presumably repair that fallibility. In so doing, however, they fail to acknowledge a far more important function of digital media in the realm of human memory. If we consider media technologies as tools for selecting, framing, and encapsulating, rather than as mechanical devices for recording and storing, then media technologies play a constitutive role in the production of memories, and hence in the continuous (re)construction of our selves (Kuhn 2000). "Technologies of the self," as Foucault (1972) calls them, involve a

For one thing, digital cameras may develop the photographer's ability to manipulate memories. Since pictures are stages in a process of remembering, their new materiality is bound to affect their status as mnemonic aids. Instead of reinforcing the modernist belief in a division between "authentic" and "manipulated" memories—a division firmly held to by MyLifeBits—digital technologies may help reconceptualize memory as a process etched in time, as a mutating sequence of objects to be worked on, continuously subject to the vagaries of reinterpretation and reordering. For another, digitization may seriously affect the gendered practices of memory production and storage. Since the ordering and storing of memories no longer participates in a different "materiality" but is now an inherent extension of digital production, the conventionally female domain may attract the interest of males. There is a historical precedent for this reshuffling in the gendered use of media technologies. As Douglas (1999) has so beautifully documented, a radio set in the 1920s required the use of quite sophisticated technical skills (commonly the province of males) before it could be made to function as a media technology in a private home, and yet the primary function of this new appliance—allowing users to listen to music—attracted more men to what used to be a typically feminine cultural practice. The new "listening box" provided a perfect vehicle for men to explore emotional and social realms, but their tinkering with the machine still affirmed their masculinity. Thus the radio, as Douglas (1999: 14) states, played a central role in "tuning and retuning certain versions of manhood" (ibid.: 14). By the same token, it may happen that the digitization of memory tools will play a pivotal role in the renegotiation of "feminine" and "masculine" traits in memory production, storage, and (re)presentation.

Multimediation of Memory

The multimediation of memory tends to erase the specificity of particular media, and yet multimediation may also allow for a new synesthetic quality of memory.

People's individual memory objects and acts are structured by the logic of singular media types, woven into specific singular practices such as taking photographs, making home movies, or recording audiotapes. Most ordinary users exhibit an unarticulated preference for one medium over another—for instance, preferring photography to the taping of moving images. The cultural forms and practices inherent in these singular media technologies unconsciously shape the recording of experiences and thus profoundly affect later acts of remembering. We may not always be aware of how the choice of one medium over another quintessentially defines the content of mediated memories, let alone of its impact on the construction of self-identity over a lifetime. The fact is, however, that the availability or coincidental presence of certain media technologies in one's life often determines one's preference for preserving memories in the form of text, audio files, or still or moving images.

How can the new multimedia equipment engender transformation in the process of remembering? It is peculiar that software project engineers take for granted people's desire to store "experiences" via exhaustive inscriptions of all their sensory facets so as to remember them "correctly" and "completely." But mediated memories usually serve not as exact recordings but as evocative frames. People want a representation that triggers particular emotions or sensations, not one that reinvokes the experience as a whole. The recording medium once dictated one's choice of sensory inscription, but the question now is the extent to which the multimedia computer will change that choice, since digital cameras, in combination with software packages, can now be promoted as recorders not just of still pictures but also of moving images, sound, and text. I don't think that people have suddenly conceived a desire to use digital equipment in recording "complete" sensory experiences, but people may begin to experiment with the synesthetic qualities of the new machines. Just as parents used to preserve a toddler's first words, her cute hairdo, and her nonfigurative artistic expressions as separate paper or taped records, digital equipment now enables postmodern parents to comprehensively document a child's world in all its visual, auditory, and textual dimensions, and in one take. The ability to capture—perhaps inadvertently—multiple dimensions may shift a person's propensity to privilege one sense over others in the process of remembering. In other words, new digital technologies, by appealing to a variety of senses, potentially change the way we choose to frame the past in multimedial modes.

Googlization of Memory

The notion of Googlization tends to emphasize the idea of memory as a process of searching a fixed database, whereas the emphasis should be on the inherent connectedness of individual memory to a constantly evolving social context. When the model of an intelligent software agent (Google) is projected onto the process of remembering, the inherently transgressive and transformative qualities of human memory are effaced in a remarkable way. Memories, in the model that projects Google onto the process of remembering, are fixed entities on the shelves of the mental library, patiently awaiting a retrieval that will be triggered by a thought, conscious or coincidental. The digital imperative to "capture," "store," and "retrieve" personal memories conceived as fixed items seems to be based on outmoded notions of memories as mnemonic imprints and of the mind as an immense warehouse. Such notions have been discredited in favor of conceptions of a complicated encoding process whereby memories are preserved through elaborate mental, social, and media schemata. These schemata rework what the minds retains, "thereby shaping memory according to shifting forms, scripts, and patterns which themselves mutate along with changing social circumstances or according to the nature of the information received" (Gross 2000: 3). The content of a memory is significantly structured by specific needs, interests, or desires on the part of the one who remembers, just as it is also structured by collective experiences and technologies. It is my contention that human memory is not a fixed object but is instead a flexible agency through which identity development, and thus personal growth, is made possible. The power of Google lies not in its ability to search fixed sets of databases but in its ability to guide a person through a vast repository of ever-mutating items, yielding different content according to when and how these items are retrieved. Two unique qualities of Google—its connectedness to the Internet, and its ability to track and reproduce continuous mutations—are notably absent from MyLifeBits's projections onto human memory.

The denial of human memory's inherent mutability may seriously hamper the high ambitions of software designers. Projects like MyLifeBits tend to regard software programs as self-contained nostalgia machines—jukeboxes of individual memory—rather than as instruments of creative production that are continuously connecting the self to the larger contexts of community, society, and history. Memory, as recent theories propose, is not exclusively cemented either in the recall of individual experiences or in that of collective experiences, but a human being has a vested interest in connecting the two if she wants to pursue personal growth. As Thompson (1995: 209) concludes in his study on media and modernity, "Self-formation has become increasingly interwoven with mediated symbolic forms, and therefore mediality is a defining factor in the construction of self as a symbolic project." Digital technologies, particularly those based on use of the Internet, have the ability to link personal memory with publicly mediated materials, thus eliciting insights into the interconnections between self and world. For example, a diary or a scrapbook, in analog form, serves as a reflective instrument within the contained universe of a person's life (Mallon 1984; Katriel and Farrell 1991), but the new potential of networked computing is not, as Gordon Bell would have it, to scan all words into the computer in order to render one's personal testimony searchable via keywords; rather, the real innovation is the computer's ability to allow a new type of diary (for instance, a weblog), a networked materiality that serves as the precondition for linking private reflection with public events, for opening up one's personal reflections to reciprocal reflections by others. The Googlization of memory is a callow conceptualization of what networked modalities can produce: true innovation in a digital memory machine would enable the emergence of new genres for connecting personal memories to others' reflections or to public resources. Needless to say, such a transformation calls for renewed awareness of the relationship between personal and collective memory; the Googlization of memory has substantially larger consequences than Gordon Bell and his associates want us to believe.[9]

Let me return, finally, to the problem of my friend who was facing the storage and retrieval of innumerable memory-infested digital files. Would he be significantly helped by a project like MyLifeBits? Would this software help him restore order to his plethora of digital memorabilia and give him time to relive the captured moments? My friend, confirming my hypothesis, told me that he was not exactly waiting for a sophisticated program or an intelligent agent to help him find particular items and turn them into smooth multimedia presentations. "The funny thing is," he said, "I am not very keen on retrieving the experiences I recorded. The value of my personal digital collection is situated first and foremost in the fun of recording and collecting, and perhaps second in knowing that these files are somehow stored, in coding, even if I will never retrieve them." His conclusion—that he treasures the act of collecting more than his actual collection—is not unique to those who record the events of their lives in the digital age; rather, it is quite analogous to the situation of a woman who keeps her love letters wrapped in ribbon in an attic but never, or almost never, looks at them. Apparently my friend's digital recordings had already served their purpose in the act of memory, even though, as objects of memory, they may never materialize beyond the virtual stage. While building his new digital environment, he had noticed how the new tools helped him arrange his everyday world and find new ways of inscribing his personal life in the dimension of time, in an orderly way.

And yet it was not simply or purely his own personal life that he wrapped in bits and bytes—it was the life of his family. His partner had protested his proposal to digitize the entire family's shoebox, and he respected her different take on preserving their common analog heritage. His technological tinkering had taught him the value of their collection as a communal resource, and her decisions about storage made him aware of their collection's preciousness. The last thing he wanted was a single technological system that would erase these idiosyncratic preferences for storage and retrieval. Acts of memory, in short, are reified in practices that are material as well as sociocultural, and these in turn affect our understanding of ourselves and of our relationships with others and with our own past and present. It is the very thing that we love and look for in the digital—its morphing, flexible capabilities—that will have the biggest impact on what our memories will become.

Memory never has been quite what we remember it to be, since its mechanics, its materiality, and its social use change along with its technologies. The performative nature of memory is, I believe, very much underemphasized in the current research on memory tools. Engineers and visionaries, in their search for the perfect digital memory machine, have focused systematically on the products of memory and ignored the role of technologies in the active staging of memory—in the mutual shaping of memory and machine. Designers of digital tools could profit from more expansive ethnographic research—inquiries into the individual's creation and use of material as well as digital memories, and into processes for collecting, storing, and retrieving them. There is already an interesting body of research on the cultural meaning of material types of collecting and storing objects for later remembering (Muensterberger 1994; Pearce 1999; Danet 1997). Building on this research, we can try to find out more about the new digital materiality and about how it affects our everyday habits of constructing a personal and family heritage. Digital technologies seem to promote a different materiality, one that both complements and partially replaces analog objects embodying memory. Most important, these technologies shape the very nature of remembering as they become (literally) incorporated into our daily routines of self-formation. Contrary to contemporary software designers' assumptions, I think the ultimate goal of memory is not to end up as a PowerPoint presentation on a grandchild's desktop. The ultimate goal of memory (and of memory machines) is to help us make sense of our lives, to create our own "living" meanings of life.

NOTES

1. "Cultural memory" is a rather broad term that I find most satisfactorily explained by Bal, Crew, and Spitzer (1999), who define it as a process that can be understood both as a cultural phenomenon and as an individual or social one. Cultural memory inherently involves the mutual shaping of individual and collective, of self and community, through various products and processes of culture.

2. Mediation of memory is a concept that

I loosely derive from Thompson (1995), who theorizes the role of media in the construction of social and personal experience. See van Dijck (2004) for further explanation of this concept, which also forms the conceptual heart of my book on the use of digital technologies in personal cultural memory (van Dijck 2007).

3. When I speak of "media technologies," I actually mean the intricate connection between technologies (hardware and software), cultural forms, and cultural practices. A photograph, for instance, encompasses the apparatus of the photographer, the cultural practice of (amateur or professional) photography, and the ensuing forms in which the picture materializes (for example, in a family album). In other words, the instruments, acts, and products of photography constitute the framework of memory.

4. See van Dijck (2005b) for a more extensive description and comprehensive comparison of various types of projects devoted to digital memory machines.

5. The scientific premises of the MyLifeBits project can be found in Jim Gemmell, Gordon Bell, Roger Lueder, Steven Drucker, and Curtis Wong, "MyLifeBits: Fulfilling the Memex Vision," downloadable from http://research.microsoft.com/barc/MediaPresence/MyLifeBits.aspx (retrieved March 17, 2007).

6. See, for instance, Scheeres (2002). See also Hissey (undated).

7. I take most quotations from "Software Aims to Put Your Life on a Disk" (2002).

8. Scheeres (2002).

9. Google's algorithms will never be instilled in human memory, but the fact that Google recently bought Blogger.com, one of the first weblog-software start-ups, indicates the company's sharp eye for the individual's potential to navigate the baffling complexity of the public and private roads crossing the digital world.

9 ▪ Tunnel Vision

Inner, Outer, and Virtual Space in Science Fiction Films and Medical Documentaries

ANNEKE SMELIK

Images of tunnels are abundant in both science fiction cinema and medical imagery. In science fiction cinema, the tunnel signifies something virtual and abstract: cyberspace or virtual reality. In medical imagery, on the contrary, it signifies something actual and concrete: the interior of a body. Yet both sets of images are equally virtual and abstract to the spectator.

In this chapter, I argue that the visualization, in science fiction films, of a spectacular ride through cyberspace is in fact a metaphorization based on the human body. At the same time, I suggest that the image of a ride through the body is informed by typical elements of science fiction cinema, such as speed and movement. Thus fictional cinema and medical visualization techniques mutually shape one another. Here, I explore two different representations of the cybertunnel in science fiction film that are based on analogies with the body—specifically, with the nervous system and the womb. Starting from cyberfilms,[1] I assess how the visually similar "ride" in two such different genres as science fiction cinema and medical science informs our imagination of cyberspace as well as our understanding of the body as shot through with technology.

INTO THE MATRIX

With remarkable tenacity, cyberfilms represent the trip into virtual reality as the passage through a tunnel, and they do so to the point where this kind of representation has quickly become a cliché. I use the metaphor of the tunnel in my reading of these images because the films show a rounded, closed-off, tunnel-like space functioning as a passage. The tunnel in this figuration can be seen as a new version of the old topos of the tunnel as a passage from one kind of world to another, such as we know it from fairy tale and myth. Computer technology enables filmmakers to create virtual tunnels with the use of digital special effects, which spectators find quite overwhelming. From *Tron* (1982), one of the first movies in which characters enter the mainframe of the computer, to such films as *Freejack* (1992), *Lawnmower Man I* (1992), *Hacker* (1995), *Johnny Mnemonic* (1995), *The Cell* (2000), and the trilogy of *The Matrix* (1999, 2003, 2004), virtual reality and the trip into it are represented in conspicuously similar ways.[2]

The trip into cyberspace is, quite obviously, imaged by computer animation. It is visualized as progress through a twisting tunnel into which the character is sucked at incredibly high speed, a passage lasting usually about one or two minutes. The accompanying sounds are eerie and nonhuman. During the ride through the tunnel, the character is not represented as an actual presence; in other words, the (virtual) camera travels through the tunnel by itself. In these scenes, the absence of any image of the actor's living body is consistent with the view that "the plasticity of the image . . . has overwhelmed the reality of the flesh and its limits" (Sobchack 2004: 50). At the end of the tunnel, the (now virtual) character arrives in an altogether different dimension where cyberspace is represented as pure mind, a cosmic power, or an altered reality. Meanwhile, in most films, the "real" body of the character is shown as having stayed behind in real life. After the character's exhilarating ride through the tunnel, a much quieter and more open virtual space allows him or her to freely move, float, or fly. Nevertheless, this space of virtual reality threatens to trap the character; he or she can disappear into it, disintegrate, fall to pieces, or go mad. Death is often imminent. Finally, the character is either liberated and returned to the real world or is lost in virtual space and dies.

While most tunnel scenes follow this pattern, a distinction can be made in how they are visualized. One tunnel looks like a cybernetic grid, whereas another is closer to an endoscopic image as we know such images from medical visualization techniques. Although the second kind of tunnel is recognizable as an image that takes the inside of the human body as its point of reference, I want to suggest that the tunnel visualized as a cybernetic grid also refers to the human body: that is, to the nervous system.

An example of the trip into cyberspace as a ride into the nervous system can be found in *Freejack*. Two characters, Alex Furlong (Emilio Estevez) and Julie Redlund (Renee Russo), are about to be transported into cyberspace. They stand in the dark, looking with some apprehension at a huge eye that has appeared before them. A wind begins to blow, and uncanny sounds of swishing and screeching are heard. Then computer-generated images take over. The huge eye contracts into a

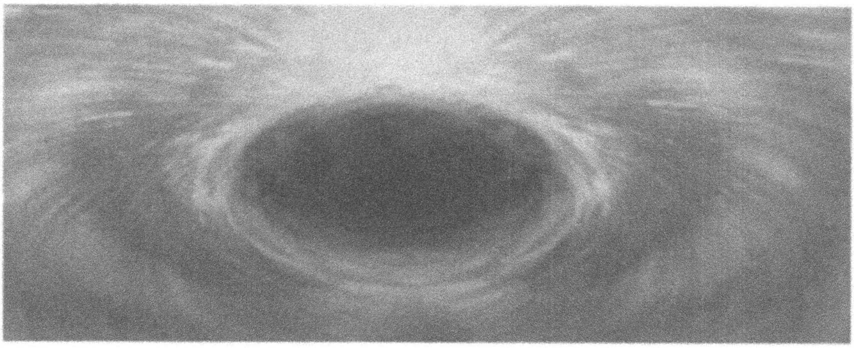

Fig. 9.1. The opening of the tunnel in Freejack

black hole (fig. 9.1), and the characters are sucked into the iris. While the swishing and screeching becomes louder, the simulated camera flies through the tunnel at an improbably high speed. The tunnel is visualized with graphic and kinetic images that flash by. After about a minute of this pure simulacrum of a ride through time and space, the camera slows down before being sucked into yet another tunnel and speeding up even more. After another minute, the camera emerges from the computer-generated tunnel, slowly descending into a rocky, alien landscape and zooming in on the stunned faces of Alex and Julie. An image of ripples resolves into the figure of Ian McCandles (Anthony Hopkins), who greets them: "Welcome to my mind."

The trip into cyberspace as a ride within the interior of the human body is similar but shows some characteristic differences, as we can see, for example, in the film *Virtuosity* (1995). "You're in my world now!" shouts one of the villains, hurling Parker Barnes (Denzel Washington) into cyberspace against his will. While Parker's worldly body is tied to the computer, his virtual body falls through a virtual city that swirls around him. His screams of fear and horror echo throughout the scene. As he falls through cyberspace, his mind is deleted, as visualized with digital special effects. A great mouth opens, and the virtual camera enters a narrow red tunnel (fig. 9.2), curving and turning while its speed and Parker's screams intensify. Farther down the tunnel, human faces appear, and again the camera is swallowed by a mouth, which gives access to yet another tunnel. The scene switches back to the computer room, where Madison Carter (Kelly Lynch) comes to Parker's

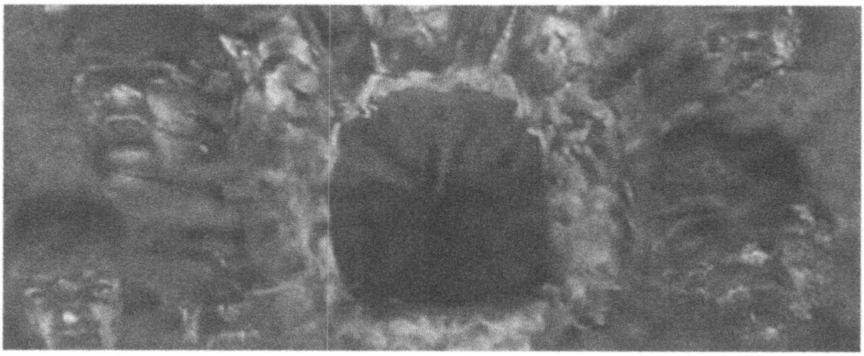

Fig. 9.2. The opening of the tunnel in Virtuosity

help by switching off the computer that is deleting his mind and his nervous system. In real life, he sits up. Surprised and in pain, he touches his head.

The main difference between the tunnel visualized as the human nervous system and the tunnel visualized as the interior of the human body has to do with the types of cybernetic images that are used. In the case of the tunnel as nervous system, the images are gridlike or chartlike, smooth, taut, and graphic. The dominant color is blue or green. This kind of tunnel is featured in *Tron, Freejack, Hacker, Johnny Mnemonic, The Cell*, and the *Matrix* trilogy. By contrast, the tunnel as endoscopic image is visualized as fleshy, bumpy, fluid, and formless, continually expanding and shrinking in every direction. The color red dominates here. This kind of tunnel is found in *Fortress* (1993), *Ghost in the Machine* (1993), *Tekwar II* (1994), and *Virtuosity. Lawnmower Man I* features both kinds of tunnels.

Nevertheless, these distinctive visual modes can be connected through a metaphor that is frequently used for cyberspace: the matrix. The word "matrix," derived from Latin and related to *mater* (mother), "matrix" originally connoted the womb or a breeding female. It is probably no coincidence that *Virtuosity* features a red, human mouth as the entrance point of the cybertunnel. In mathematics, the term "matrix" refers to a rectangular array of quantities or expressions, or to a gridlike array of elements, especially of data items. We find this kind of representation in the images of the tunnel as a nervous system. Because the nervous system is connected more with the brain than with the womb, the eye features as the entry into the cybertunnel in *Freejack*. Here, science fiction cinema borrows extensively from neuroscience and its insight that the act of representation takes place within the brain through the response patterns of sets of neurons, interpreted as "points in phase spaces," which have a metric and are visible in figures of sensory and motor grids (Churchland Smith 1986: 452–53). According to this view, cognition can be understood geometrically, as "characterized in terms of phase spaces, vectors, and matrices" (ibid.: 457). The term "matrix," then, brings together the two different visualizations of the ride into cyberspace. I will come back to this metaphor in connection with my analysis of the tunnel as a trip into the "wetware" of the human body.

THE TUNNEL AS NERVOUS SYSTEM

The image of tunneling into cyberspace conceived as a nervous system is based on an analogy between the computer and the human brain. This is typical of cyberculture, which is all too happy to postulate a sustained analogy between the computer network and the human nervous system. Scientists who have had a huge impact on cyberculture, such as Hans Moravec and Marvin Minsky, believe that in the future it will be possible to plug a computer or implant a chip directly into the human nervous system or the brain (Dery 1996), whether to download or upload information. Equations are made between computer software and the brain, between computer hardware and the body, and between the Internet and the mind. These equations are founded on certain strands in neurophilosophy that harbor materialist views of man as a machine, and especially of the brain as a machine—for example, Churchland (1984: 113, 120), who seems to place almost no distance between himself and "functionalist AI [artificial intelligence] theorists" when he claims, "If machines do come to simulate all of our internal cognitive activities, to the last computational detail, to deny them the status of genuine persons would be nothing but a new form of racism." Such outrageous claims point, at best, to a reductionist view of the human body and subjectivity and, at worst, to an irresponsible and apolitical attitude toward the role of technology, an attitude perhaps best described in the words of Virilio (1998: 182) as representing a form of "delirious technical fundamentalism." Churchland does not work from the standpoint of cyborg complexity, as it has been advocated by feminist scholars like Donna Haraway, but from what Bennett and Hacker (2003: 355, 366) describe as "ontological reductionism," the belief that all human behavior can be explained by neuroscientific theory, if not yet, then in the not too distant future.

Let us not forget that the presupposed equivalence between the structure of the brain and that of the computer is, as Hayles (1999) reminds us, fundamentally metaphorical. The analogy sustains a false binary opposition between mind and body, or between information and materiality, where the mind is understood as not being embodied, and the body is understood as being mere material flesh. In this context, mind always rules over matter. Moreover, there is continuous slippage between the mind as something spiritual (or virtual) and as something more physical that is situated in the human brain. In neurophilosophy, interestingly, this slippage is willfully turned into an equation (mind = brain) both by Churchland (1984) and by Churchland Smith (1986). Whereas Churchland, however, subscribes to the metaphor of brain as computer, Churchland Smith (1986: 459) dismisses that metaphor as "fundamentally wrong" and "a trifle thin," especially in relation to the function of memory. Thus the metaphor of brain as computer is contested in neuroscience, especially in more recent evaluations of the field, where the metaphor hardly figures at all. For example, Bennett and Hacker (2003: 432) discuss the computer-as-brain metaphor only in an appendix, where they convincingly argue that such metaphors "are poor ones" and are even "a bizarre suggestion," mostly because the computer-as-brain metaphor does not allow for enough complexity and holds on to a misconception of the self as an immaterial substance.

By contrast, the popular genre of science fiction has not been too much hindered by such theoretical niceties. Cyberfilms enthusiastically embrace the analogy between mind and computer. In *Johnny Mnemonic*, for example, the hero, Johnny (Keanu Reeves), downloads data into his brain, temporarily deleting part of his personal memory. In *Virtuosity*, a person's mind, when plugged into a computer, can be either enhanced or distorted. In *The Cell*, people have the scary ability to enter into one another's brains and memories. In *The Matrix*, Neo (again Keanu Reeves) inserts a diskette and, in just a few minutes, learns Eastern ways of combat. These are all neat, almost bodiless, man-machine exchanges.[3]

Here we touch on one of the main characteristics of cyberculture: its celebration of disembodiment and immateriality, as many critics have aptly remarked (Bukatman 1993; Dery 1996; Hayles 1999; Sobchack 2004). In films dealing with virtual reality, characters are temporarily liberated from the constraints of the body and are therefore freely able to cross time and space. The design of the images—for example, the disembodied point of view from inside the tunnel, or the transparency of bodies in virtual space—makes it possible to deduce a strong desire to transgress physical boundaries, or even to leave the material body behind altogether. The tunnel ride suggests the euphoria of becoming pure mind. In this respect, it is striking how clean the representations of such trips are: the character, free of the body, falls, flies, or floats virtually through space. These films are keen to avoid any carnality in virtual reality. The virtual apparently excludes the material "wetware" of the body. The spectacular virtual-reality fights in the *Matrix* trilogy are the most extravagant example of this utter freedom from physical constraints.

Thus the trip into virtual reality signals a flight from the flesh. In the pop culture of cyberpunk and science fiction, the posthuman subject is rapidly dematerializing. The desire to transcend the body resounds time and again in utopias concerned with information and communication technologies, which give shape and content to such concepts as cyberspace and virtual reality. Cyberfilms envisage virtual reality as one way to actualize the desired state of immateriality and disembodiment.

MEATSPACE

In scenes of virtual reality, the fantasy of the ultimate trip is caught in the Western duality of body and mind. In a double shift, the mind is first split from the body, as if the mind were not itself embodied, and then it is valued much more highly than the flesh.[4] The extent of the negative attitude toward the body can be grasped from texts produced in a cybercultural context. Cybertexts, whether fictional or theoretical, refer to the body as a living hell from which one can be liberated through virtual reality. Gibson (1984: 6), for example, writes in his novel *Neuromancer* about the "bodiless exultation of cyberspace" from which the character Case falls back into "the *prison* of his own flesh." Heim (1991: 64) writes, "Suspended in computer space, the cybernaut leaves the *prison* of the body and emerges in a world of digital sensation." And John Barlow believes that "the Net is somehow going to free us from the *tyranny* of the body . . . ; by going digital we can break free of the *prison* of the flesh" (cited in Zaleski 1997: 35; emphasis added). In this context, the euphoria of cyberspace is opposed to the pain and suffering of the body in ruins in "meatspace."

It is important to draw attention to the gender dynamics involved in this negative view of the body in cyberculture. Apart from the fact that I don't share this disgust for the body—after all, the body can also be a source of multiple pleasures—its feminine connotations are denied and repressed. As critics have argued, contrary to dominant representations in cyberpunk literature or cyberfilms, the posthuman subject is always already embodied, and its relation to gender is often problematic (Hayles 1999; Braidotti 2002). Balsamo (1996: 123) writes, "From a feminist perspective, it is clear that the repression of the material body belies a gender bias in the supposedly disembodied (and gender-free) world of virtual reality."

Because the physical, material flesh is depicted as a female body, the desire to transcend that body points to a negative view of femininity. Therefore, in the discourses of cyberculture, the desired transcendence of the body in virtual reality can be read as a flight not only from the flesh but also from femininity.

THE TUNNEL AS WOMB

The flight from femininity is complicated, however, by contemporary culture's preoccupation with visually representing space, whether that space is real or virtual, physical or cosmic. Haraway (1992: 305) has offered a reading of science through a map, or "semiotic square," in which she presents four regions of analysis: "A" (real space), "not-A" (virtual space), "B" (outer space), and "not-B"(inner space). I want to suggest that in current visual culture these different kinds of spaces are collapsed. Because of virtual space's very virtuality, and hence because of its uncertain parameters and its fundamental ambiguity, it may be open to possible confusion with real space, outer space, and inner space. For one thing, science fiction films have shifted the conventional concern with outer space to a concern with virtual space, often conflating the two in the process. For another, the inability to distinguish between real space and virtual space is very much the theme of films on virtual reality, such as the *Matrix* trilogy or *eXistenZ*. Nevertheless, what concerns me more specifically in the image of the "tunnel" is the collapse between virtual space and inner space. As mentioned at the beginning of this chapter, the trip into virtual reality is shown as a passage through a rounded, closed-off space that looks like a tunnel. Images like these are familiar to us from such medical visualization techniques as endoscopy, although science fiction films and medical visualization techniques display these images at very different speeds. Thus the trip into cyberspace is, visually speaking, uncannily similar to a trip into the human body. The images of the eye in *Freejack* (and in *The Cell*) and of the mouth in *Virtuosity*, as points of entry into the tunnel, underscore this metaphor.

Feminist scientists have shown how the scientific urge to unravel the secrets of nature is predicated on a gendered perspective whereby the male mind penetrates the inner mysteries of nature, inevitably depicted as female. Keller (1985: 31), who has traced this prevailing gendered perspective through the history of science, writes: "Laid bare of her [Nature's] protective covering, exposed and penetrated even in her 'innermost chambers,' she is stripped of her power. Her secrets have become knowable." While Keller's groundbreaking work exposed the highly gendered, sexualized metaphors of science, other feminist critics have been keen to expose similar practices in medicine, especially where the new reproductive technologies are concerned. Imaging technologies capable of probing inside the womb and making the inside of the female body visible have been at the heart of public debates about the new reproductive technologies (for example, "test tube babies") as well as about abortion. Feminist activists and scholars have been active in these debate for the past two decades. Many have performed critical analyses of the power of visualization techniques and their implications for the politics of reproduction in a culture that quickly became fascinated with fetal images (Petchesky 1987; Franklin 1991; Hartouni 1992; Stabile 1992). Other feminist scholars have been concerned with how imaging technologies inscribe science, at women's expense (Treichler, Cartwright, and Penley 1998). The female body is imaged as a mere vessel for the fetus, which gains autonomy while the mother is obliterated in a visual act of disappearance. She has become, as it were, a tunnel through which the fetus travels.

The feminist critique of science and medicine can be set against the background of feminist philosophy in the 1970s. Luce Irigaray is critical of a Western tradition in which the speculum, a concave mirror, is believed to expose the secret of female sexuality to the male subject; the instrument may allow the eye to penetrate the interior, but it produces "anamorphoses by the conjunction of curvatures" and "impossible reflected images, maddening reflections, parodic transformations" (Irigaray 1985: 144). Female sexuality will remain forever a mystery. For Irigaray, the speculum is a metaphor for the production of female subjectivity as fundamentally "Other" and "interior."[5]

It seems that some documentary makers have taken feminist criticisms seriously. As Bryld and Lykke show in chapter 6 of this volume, a U.S. version of the Swedish documentary *The Miracle of Love* altered significant parts of the original narrative, making the story less romantic and less male-dominated. The same can be said of the famous BBC documentary series *The Human Body: The Incredible Journey from Birth to Death* (1998),[6] which takes great care to give the mother-to-be full narrative and visual space in the episode on conception and pregnancy. Thus the documentary tries to avoid objectifying the woman, instead making her a subject in her own story of life. It also reverses the old sperm-meets-egg story at the microscopic level, telling instead a story similar to that of the Egg Queen (see chapter 6).

The Human Body is an excellent showcase for the latest medical visualization techniques. The series presents a "journey" through life in seven episodes of fifty minutes each, narrated by Robert Winston. I find the series most extraordinary for its bold, intimate, spiritual portrait of death in a culture that increasingly fears the process of ageing and dying, but the fame of the series is due to the novelty of many of its visualization techniques (and, of course, to the engaging personality of Winston). Indeed, Winston often refers with excitement to showing something "for the first time," or to things "never before seen on television," or to taking the camera to "places where it has never been before." And the DVD of the series, following the custom established with movies, comes with an extra DVD that deals with "the making of" *The Human Body*, a series eminently proud of its scientific innovations, its visual technologies, and its computer graphics.

What concerns me here is how so many of the visualization techniques used in the series produce the image of a tunnel by way of an endoscope (fig. 9.3), described as a "minitelescope," moving through the bowel, going into the eye or the ear, exploring the womb, traveling along the penis or an artery or a gland, or following the optic nerve right through the brain. Other techniques and tools that appear in the series are scanning electron microscopy, magnetic resonance imaging, ultrasound imaging, time-lapse photography, time-slice camera work (involving 120 cameras on a frame), infrared cameras, and heat-sensitive cameras. The images are enhanced by computer graphics, sometimes involving a crew of dozens of people working for three days or for as long as several months. Each episode of *The Human Body* features between three and five tunnel images, with a total of some twenty-five in the series as a whole giving views of the interior of the body (fig. 9.4). Some of these images are virtual, in the sense that they were produced by digital animation, and some are real, that is, based on endoscopy or other visualization techniques.[7]

In both cases, the imagery is often, if not always, remarkably abstract and "unreal," a quality that places this imagery visually close to the computer-generated images of the cybertunnel in science fiction films. Many microscopic, endoscopic, and other scopic images yield disturbing, abject imagery of which a science fiction or horror movie would be proud. In fact, the images cannot be read by a medically illiterate spectator—that is to say, by most of us. They must be fully explained by the narrator and visually enhanced with computer graphics. But contemporary spectators of science fiction movies are already familiar with images of the tunnel as a passage through space, whether real or virtual, inner or outer.

An important aspect of tunnel imagery is its function of representing a passage, which indicates movement. In "real" life, the tunnel is a means of transporting a vehicle from one place to another; in a cyberfilm, a digital tunnel transports a character from earth to cyberspace. In a medical documentary, the "tunnel" is not so much a means of transport as a vessel or tube placed within the human body and traversed by a camera (or another visualization tool) for the spectator's "edutainment"—not unlike what happens in a nature documentary that charts foreign territory. The

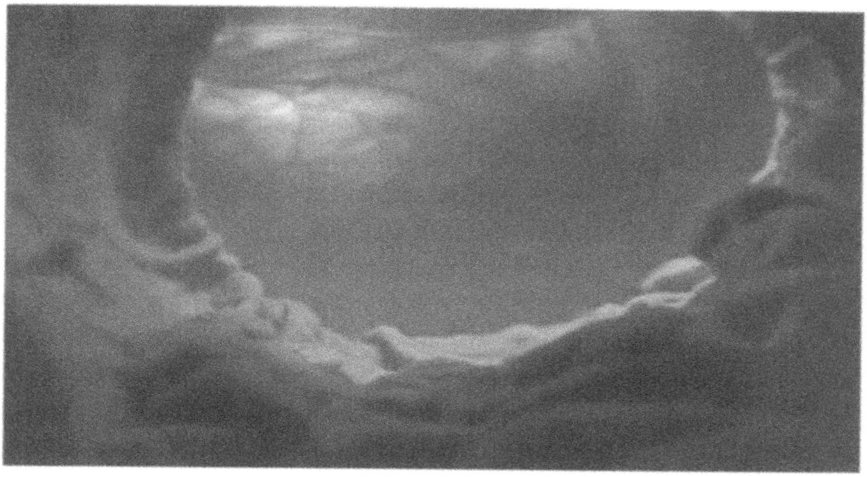

Fig. 9.3. Endoscopy in The Human Body

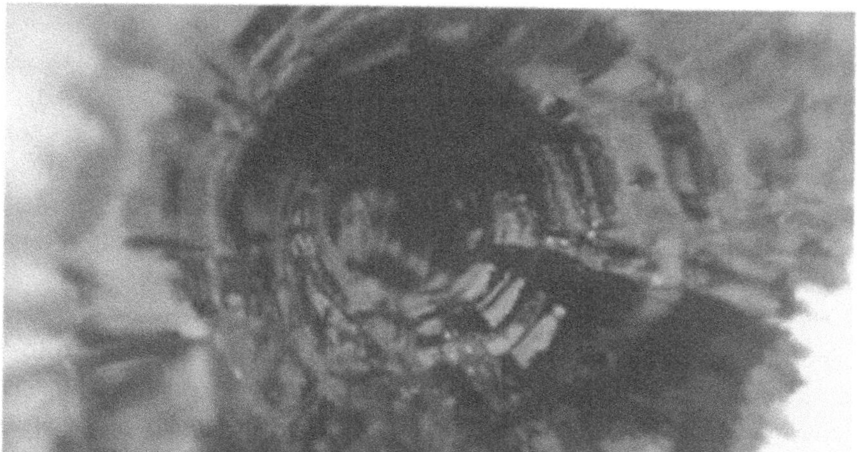

Fig. 9.4. An abstract "tunnel" in The Human Body

tunnel metaphor used in this chapter is closely related to a notion introduced by Sawchuk (2000: 14), for whom the ever-expanding field of visual media that offer anatomical images as entertainment points to a "biotourism" by which the inner body is spatialized and made into a "bioscape," in an act that links the body with geography. *The Human Body* fits in very well with the idea of biotourism; in fact, Winston often uses such travel-related terms as "journey," "trip," and "ride," as in comments about the "journey through life," or about "muscular contractions [that] propel [sperm] on its final ride through the man's body and into the woman's." He also regularly visits remote places on earth in order to demonstrate certain aspects of the human body, in visual mimicry of the technical journey inside the body. At one point, he even turns up on a roller coaster at an amusement park, as a metaphor for the "roller-coaster experience of puberty," a period when hormones are responsible for enormous changes in the body. When the roller coaster enters a dark tunnel, the documentary cuts to a tunnel inside the body, not specifying what it is showing but confirming the conflation between inner and outer space.

As feminist critics have pointed out, many medical visualization techniques were initially developed to open the womb up for inspection. Pictures of the inside of the womb have become part and parcel of Western popular culture, not least because of the books and films of Lennart Nilsson (see chapter 6, this volume). In the popular mind, then, images of the inside of the human body are first and foremost images of the womb. As a result, spectators may consciously or unconsciously transfer the image of the womb onto any endoscopic image of any other organ. I, for example, remember being struck by the similarity between medical visualizations of the womb and endoscopic images of the stomach when I saw the Australian cyber-artist Stelarc present his *Stomach Sculpture* (Stelarc 2000; Zylinska 2002). Medical visualization techniques demonstrate the extent to which the interior of any body is made up of "empty" space.[8]

Baert (2001) has argued that space and the body alike are figures of the feminine. The fact that the inner space of the body can be traversed, and visualized, doubly feminizes the human body. Its secrets have now become knowable: not only is the interior of the human body made visible, its interior consists mostly of cavities. Thus medical visualization techniques defy one of the defining characteristics of sexual difference in the biosciences: the notion of the female body as an empty vessel; the documentary shows that the male body is equally vacuous. This is not to say that medical documentaries don't continue to foreground sexual difference; which they do, both in their narrative structure and in their commentary; rather, it is to say that visual techniques belie and undermine linguistic forms of narration.

Not only have feminist critics turned their attention to medical visualization techniques, feminist artists have also used these techniques to question issues of embodiment. In Wendy Kirkup's film *Echo* (2000), for example, Kirkup takes more than twenty minutes to "map" her own body through ultrasound imaging technology. This visualization is a welcome break from the short "rides" offered in cyberfilms and medical documentaries because it takes time to fully explore the body in its complex layering. In taking this approach, the artist refuses to subject her body to the fragmentation of medical imagery or the disembodiment of cyberfilms. Yet Kirkup does not present a holistic view of the feminine body, because the exposure of its "bits of life" is still fully mediated by visual and medical technologies. Moreover, the art project that used this film consisted of simultaneous projections of the film, via satellite, onto the facades of buildings in Glasgow and Newcastle, to conflate the boundaries between inner and outer space, body and city, private and public. As Baert (2001: 59) argues, the visually and aurally striking artwork called *Echo* was "an intervention into the cultural imaginary of the female body and into powerful 'masculine' technologies," one that reworked the ways in which medical documentaries treat sexual difference. The film *Echo* traces the processes of materialization and visualization of the human body, producing complex images of technocorporeality.

FLIGHT FROM THE FLESH

So far, I have been arguing for the topos of the tunnel as a dominant representation of virtual or inner space. I have traced two sets of visual similarities: one, in science fiction films, between the tunnel as used in trips into cyberspace and as used in trips through the human body; and another, in medical visualization techniques, between images of the womb and images of other internal organs. If we bring these two sets of similarities together, it follows that the tunnel in cyberfilms can be read as affording a ride into the womb, whereas the trip through the human body can be read as a ride into virtual reality. In other words, inner and virtual space are collapsed in contemporary visual culture, which also implies that inner and outer space are no longer separate, and that the real cannot be distinguished from the virtual. We are confronted with a visualization of cyberspace that is based on the inner space of the maternal body, in a pop technoculture that privileges disembodiment. Popular culture's hijacking of this theme indicates a break between science's material "bits of life" and their immaterial representation in science fiction. How are we to understand this contradiction? I want to suggest that this paradox may indicate, as it were, a "return of the repressed."

The exhilarating ride into cyberspace can be understood as a desire to unite with the matrix, as Springer has argued. Here, the original meaning of the word "matrix" (womb) comes to mind once again. Read in a Freudian light, this return to the matrix is quite ambivalent, a "simultaneous attraction and dread evoked by the womb" (Springer 1996: 59). The fantasy of getting rid of the body, of sheer weightlessness, is often accompanied by dissolution of the consciousness in virtual reality. Through the human fusion with electronics, consciousness melts into the matrix of cyberspace— a fusion that "popular culture frequently represents as a pleasurable experience" (ibid.: 58). In a later article, Springer (1999) returns to the idea of cyberspace as maternal substitute, understanding cyberspace to be a fantasy of the mother's plenitude. Although I support this view, I think that the notion of the matrix can be expanded if it is read through Irigaray (1985). Plant (1995, 1997), referring to Irigaray's ideas, has explored some of the feminist implications of cyberspace as matrix. She sketches the age-old dream of transcendence as a male desire to be redeemed from physical matter, a desire interpreted by Irigaray as a flight from the mother. Plant refers to the metaphor of the matrix as womb but does not explore it further. Therefore, let me return to Irigaray.

In her exposition on Plato's cave, Irigaray (1985) refers throughout to the womb as the matrix.[9] The cave metaphorically signifies the origin of life, the maternal-feminine. In order to shed his chains and step out of the cave into light and reason, the male prisoner must literally turn his back on his maternal-material origins. Irigaray criticizes the male tendency to forget and repress those origins. According to her, undoing this forgetting requires an enormous revolution; to actually remember, let alone value, these origins is a nearly impossible feat. The male prisoner pays rather a high price for leaving the cave and acquiring his freedom from the maternal-feminine: he is not allowed to remember the cave, let alone nurture any desire to return to it. Thus loss of maternal origins implies first of all a loss of memory. Because the maternal and the material are always intertwined, this original loss also entails a painful repression of the enfleshed body. This is not without its consequences. Irigaray (ibid.: 273) points out that the male prisoner, once freed, suffers "dizziness, dazzlement" and even "aphasia."

I want to argue that the trip into virtual reality can be read as the desire to return to maternal-feminine origins. That desire is by its very nature both subconscious and ambivalent. The male character wants to transcend the constraints of his body, that wet, abject meat. Hurled into cyberspace, he experiences the momentary euphoria of becoming pure mind. At the same time, cyberspace in all its unpredictable fluidity reminds the character of that which he was supposed to forget forever: the womb of origin. The character then panics and becomes desperate to set himself free once again from the maternal-feminine, and to return to his own body and his own world.

The despair of the male character when he realizes that he is trapped in a maternal-feminine space can be seen in the tortuous tunnels, which are indeed both dizzying and dazzling. In that maternal-feminine cyberspace, men are no longer able to speak but cry out in jubilation (Jobe in *Lawnmower Man I*) or pain and fear (Alex in *Freejack*, during the mind transplant). It is significant that all tunnel scenes are without dialogue. Screams are the only human sound we hear. Consider the tunnel ride in *Virtuosity*, described earlier: Parker screams while his mind and nervous system are being deleted, and the accompanying sound effects make his screams even more frightening. Aphasia, indeed.

Classical cinema is predicated on the nonsignifying female cry (Silverman 1988); in genres like science fiction and horror, however, men are increasingly the ones who cry out in fear or pain. These films unwittingly provide insights into the dangers of forgetting the body, of denying the flesh, of leaving the feminine behind. The return of the repressed is visualized in sinuous, tortuous, vertiginous tunnels, as indicators of the maternal-feminine. The flesh demands its due; the tunnel threatens to swallow the character. Again, *Virtuosity* serves as a point of reference here, although the digitized mise-en-scène holds for many of the other tunnel scenes as well. The turning and twisting imagery, together with the accelerating speed and the eerie sound effects, make the tunnel into a surging vortex, where the character hovers between euphoria and agony, between delirium and terror. The desire to lose the self by entering the matrix and leaving the body behind is a fundamentally ambivalent one because the original mother-womb not only gives life but also takes it away. The cyberfantasy is simultaneously claustrophobic and ecstatic, pleasurable and painful. Alongside the euphoria and utopia of liberation from the body we see fear and dread of the body's or the mind's final demise.

In my reading of science fiction films, the material groundings of the body are never really lost, in spite of cyberculture's deployment of special-effects overdoses to market an artificial technobody. The least I can claim here is the living body's simultaneous erasure from and inscription in contemporary technoculture.

CONCLUSION

In this chapter, I have traced a particular visualization that has become a trope in current visual culture: a tunnel ride through a confined space. In science fiction films, this space is an outer, virtual one, an indication of cyberspace. In medical documentaries, such images represent the interior of the human body. Both genres feed on each other, and the result is, on the one hand, a metaphorization of cyberspace as the inside of a human mind or body in cyberfilms and, on the other hand, the virtualization of the inner space of the human body in medical documentaries.

This mutual exchange of imagery and technology points to three issues. First, it reveals the extent to which the human body has been reconfigured and remediated in contemporary visual culture. The proliferation of discourses surrounding the body has been well documented in cultural studies, feminist studies, and science and technology studies (this volume testifies to that development). Images of the tunnel, as I have analyzed them in this chapter, indicate many conflicting ideas and desires regarding the human body. The desire for disembodiment, so manifest in cyberculture, is first undermined by the equation of the computer with the mind, the brain, or the nervous system. It is further undermined by the image of cyberspace as the inner space of the maternal body. Both metaphors, summed up by brain and womb, indicate a return of carnality at the very heart of a cyberculture that desperately wants to rid itself of human physicality. Moreover, the proliferating visualization techniques that open the human body up to control, scrutiny, and surveillance point to an interest in the body that borders on morbid fascination.

Second, the desire for disembodiment and the proliferation of images of the body are two sides of the same coin in a culture where inner and outer space, not to mention real and virtual space, can hardly be told apart anymore. Inside out and outside in are collapsed in an imaginary space, which is ambiguous to the extent that the real and the virtual are thoroughly confused. Science fiction films have shifted their focus from outer space to virtual space, often conflating the two in the process, but they also foreground the failure of the characters, and sometimes even the audience, to distinguish between real space and virtual space. Moreover, as the images found in medical documentaries and science fiction films feed increasingly on one another, virtual space and inner space become visually analogous and equivalent. Whereas the semiotic square of Haraway (1992) carefully distinguishes among inner, outer, real, and virtual space, I want to argue that in contemporary visual culture this square has become a Gordian knot.

Third, in a visual culture dominated by digital technology, visualization becomes more and more a synonym for virtualization. Virilio (1998: 115) has pointed out that the conquest of space goes hand in hand with the conquest of the image. In this respect, it is quite remarkable that science fiction films and medical documentaries alike are seeking, in the image of the tunnel, to conquer a particular space:

cyberspace, or the interior of the human body. The conquest of that "final frontier" is purely visual, in the sense that it is a quest for visualizing spaces—inner, outer, real, virtual. Space is there to be rendered as an image—in the particular cases that I have highlighted, as an image that looks like a tunnel. The tunnel enables the visualization of space as a passage. Space, whether body space or cyberspace, apparently can be shown only as something to be crossed, routed, and traversed. Thus it is movement that makes space "visualizable." Not only does digital imagery conflate different kinds of spaces, as I have argued throughout this chapter, but immense speed also collapses time and space. In that sense, the visual is always in the process of becoming the virtual. It may not be an exaggeration, however, to claim that the opposite is also true: the virtual is always in the process of becoming the visual.

NOTES

1. I use the term "cyberfilms" to designate a subgenre of science fiction movies that take the world of cyberspace and virtual reality as their theme. In this chapter, I treat the terms "cyberspace" and "virtual reality" as synonyms.

2. In addition to the films just mentioned, I have found this specific representation of virtual reality in *Brainstorm* (1983), *Fortress* (1993), *Ghost in the Machine* (1993), *Tekwar II* (1994), and *Virtuosity* (1995). The abundance of tube images in *What The Bleep Do We Know?* (2004) points to stereotyped New Age representations of time travel. The iconography of the tunnel is not entirely new to science fiction films; the best-known and earliest tunnel is undoubtedly the one featured in the psychedelic trip at the end of *2001, A Space Odyssey* (1968), which set the standard for the visualization of space travel for a long time to come. *Contact* (1997) features an extended trip into the cosmos as a ride through a tunnel at breathtaking speed. Claustrophobic tunnel-like constructions can also be seen in the spaceships of the *Alien* films. Virtual tunnels are found at times outside science fiction—for example, in *Being John Malkovich* (1999). By now, the cybernetic image of the tunnel has become so standardized in visual culture that Windows Media Player, included with Microsoft XP and other versions of the Windows operating system, offers several such images as accompaniments to music played on a personal computer. The tunnel is also a recurring visual theme in MTV's "Idents," short films or animations linking television programs and featuring the channel's logo.

3. Only the film *eXistenZ* (1999) escapes the mind-body binarism by focusing on physicality: the porthole in the spine where a game pod can be plugged in. Also, instead of enhancing the human body with technology, the film turns this proposition around: technology is made of flesh, to the point of absurdity. In this respect, the film is more loyal to the genre of horror than to that of cyberpunk.

4. The Canadian director David Cronenberg is one of only a few filmmakers who

explore fear of and disgust at the feminization of men, by visually lingering on penetrations of the male body—for example, the vaginalike opening up of Max Renn in *Videodrome* (1983), or the anuslike "bioport" in the body of Ted Pikul in *eXistenZ*.

5. Cheetham and Harvey (2002), in their exploration of the cave as a trope in contemporary art and other cultural discourses, also refer to Irigaray's reading of Plato's cave as a gendered metaphor.

6. Produced and directed by Richard Dale and others, and presented by Robert Winston, this 1998 series won many prizes. The DVD was released in 2001. In 2000, the BBC produced the series *Superhuman: The Awesome Power Within*, also narrated by Winston.

7. For an excellent critique of the impact of medical visualization techniques and their production of the "transparent body," see van Dijck (2005b).

8. I place "empty" in quotation marks because many organs need to be emptied of their contents before a camera can travel through them, or before they can be seen and imaged through other techniques.

9. The link between virtual reality and Plato's cave has often been noted; see, for example, Cavallaro (2000: 28) on the development of the "cave" in virtual technology. Cavallaro describes the term "cave" as "redolent of Plato's philosophy," referring as it does to "a system that uses a pointer, worn by a guide for a small group of people, in a dome wherein virtual images are projected." The notions of simulacrum and projection also play an important role; for Hillis (1999: 137), virtual technologies "participate in a metaphysics of light as old as Plato's cave."

10 ▪ What If Frankenstein('s Monster) Was a Girl?

Reproduction and Subjectivity in the Digital Age

JENNY SUNDÉN

Female machines are a rare species. And when they do occur—from the classic example of the false Maria in Fritz Lang's 1927 film *Metropolis* to the dangerously high-heeled, slick red-leather terminatrix in 2003's *Terminator 3*—they tend to be rather explosive incarnations of sexual danger with disastrous cultural implications. The fear of machines that become uncontrollable is entwined with the fear of female sexuality that gets out of hand (Huyssen 1981). As a representation of such male anxiety, the woman machine must always "die" at the end of the story—or at least be terminated. Therefore, it is necessary to imagine a different scene. What if she does not die? What if the female machine simply refuses to die where the narrative ends? Or what if she even objects to narratives that have endings?

Instead of maintaining an opposition between disembodied, high-tech masculinity and embodied, earthbound femininity, this chapter engages in a discussion of intimate couplings between women and machines, insisting on an understanding of the subject that exceeds the poststructuralist "text," in the direction of a materialistic approach of embodied sexual specificity. Female machines, by clearly illustrating that the connection between masculinity and machinery is not given by nature but is something that constantly needs to be constructed, not only denaturalize the alignment of woman with nature but also prove impossible the very division that produced this association in the first place.[1]

One fascinating example of a female machine that refuses narratives with endings is Shelley Jackson's already classic, provocative, witty hypertext work *Patchwork Girl* (1995). *Patchwork Girl* is CD-ROM-driven hypertext fiction composed in the Storyspace authoring software. In interlinked spaces of texts and images on a computer screen, it tells alternative stories of a female Frankenstein's monster.

This chapter sets out to explore stories of artificial-(wo)man making. Letting a reading of Mary Shelley's *Frankenstein* that pays attention to the significance of monstrous sexual specificity serve as the backdrop, the discussion will turn to Jackson's change of medium as well as of the sex(ual identity) of the monster in her "remediation" (Bolter and Grusin 2000) of Shelley's tale. How does this shift in writing technologies, as well as in the sexual specificity of the monster's body, change the understanding of the (writing) subject? Like Shelley's *Frankenstein*, Jackson's *Patchwork Girl* raises questions not only about the limits of bodies on the cultural margins but about the very limits of humanness, of life itself. What does it mean to be made, not born? Or, as Stacey puts it in this volume (chapter 7), "How can we recognize authenticity and individuality in . . . the culture of the copy?"

THE SEXUAL SPECIFICITY OF (WO)MAN MAKING

In *Frankenstein*, the monster originates from a "workshop of filthy creation" where the mad scientist Victor Frankenstein organizes matter from "the dissection room and the slaughterhouse" (Shelley [1818] 1994: 52). In isolation and guilt, he learns the secrets of the human frame by observing the processes of corporeal decomposition. Human life is created by way of decay, death, and galvanism, and the result is a monster of the most hideous proportions. The monster's tragedy is his solitude, his exclusion from human relations in general and from an intimate connection with a female counterpart in particular. He begs Victor to create a female monster, but halfway through her construction Victor decides to destroy her:

> I thought with a sensation of madness of my promise of creating another like to him, and trembling with passion, tore to pieces the thing on which I was engaged. The wretch saw me destroy the creature on whose future existence he depended for happiness, and with a howl of devilish despair and revenge, withdrew [Shelley (1818) 1994: 161].

While the monster vengefully watches him through the window, Victor violently terminates his second act of (pro)creation, since this female monster may "become ten thousand times more malignant than her mate" and may even "become a thinking and reasoning animal." The monster has promised to hide from human society, but *she* has not. There is no guarantee that she will voluntarily adapt to the subordination of femininity. Moreover, the male monster may hate her deformed body more than he hates his own, or, perhaps even worse, she may "turn with disgust from [the monster] to the superior beauty of man" (Shelley [1818] 1994: 160).

This fear is a central theme in James Whale's 1935 horror movie *The Bride of Frankenstein*. What if female monsters desire men instead of monsters? Or women? Or each other? Jacobus (1982) points out that the fantasy of the female monster who longs for men instead of monsters is a terrifying threat to male (hetero)sexuality. This threat is tied not only to the she-monster's hideous deformity but also to the frightening autonomy of her refusal to reproduce in the image in which she was made.

Feminists reading *Frankenstein* have often made sense of Victor's creation of the monster, not in terms of the making of a man by another man, but in terms of the more familiar framework of female sexuality, reproduction, and, ultimately, abortion. *Frankenstein* is regularly referred to as a horror story of maternity, as a symptom of Shelley's own difficult experiences of giving birth, a reading that turns Victor into a fallen woman with an illegitimate child (Donawerth 1997; Gilbert and Gubar 1979; Moers 1977; Veeder 1986; Youngquist 1991). This type of reading can be challenged on at least two grounds. First, while such a reading productively puts forth the moral responsibility of the creator for his own creations (a responsibility that traditionally has been associated with women), it makes the life of the woman author the primary source of understanding, reducing literature to a mere reflection of "her" reality. Second, to read *Frankenstein* as a maternity story is to fail to take account of the radical element of artificial-man making and its attempt to bypass the maternal.[2] Time and again, as if there were some shortage of cultural imagery to account for the creation of life without a woman, the making of Frankenstein's monster is collapsed into a more intelligible framework of reproduction—that of motherhood.

Readings of *Frankenstein* along the lines of motherhood miss the point that the creation of the monster depends on the absence, or even death, of the mother (Homans 1986). Frankenstein's reanimation of a dismembered dead body introduces a way of creating life that no longer needs a mother—and, in fact, all the mothers who appear in the course of the narrative end up dead. This act of "unnatural" procreation fundamentally changes the meaning and function of family relations, and in particular of heterosexual relations. At least one explanation for the renewed attention being brought to *Frankenstein*, in an era when biotechnology and artificial life are becoming commonplace, is the novel's picturing of alternative ways of creating life, of single parenthood and the lives of illegitimate offspring.

Then again, playing God comes at a price. Victor's monstrous incarnation of a male fantasy of autonomy and creativity returns with a vengeance, showing him that issues of life are always deeply entangled with those of death, and that a creator can never fully control that which he has created. Even if the monster has no understanding of death, he himself is an embodiment of it. To simulate life mechanically is to accentuate human mortality—to invite a reminder that being human means being mortal. In a narrative thread that parallels the dream of cyberspace as a high-tech version of masculine bodily transcendence (see chapter 9, this volume), Victor attempts to eliminate female sexuality in favor of scientific principles that supposedly purify the previously impure act of procreation. The terrifying outcome is a grotesque version of himself, a distorted copy (but nevertheless a copy) that proves to him that the force of life is not divine but can be technologically manipulated (Waldby 2002).

The monster's being male is important for this mirroring process.[3] Victor's hatred of the monster derives from his having made the monster in his own image; hatred of the monster is reflected back on him as self-hatred, as an inability to acknowledge and take responsibility for the monstrous and mechanical dimensions of human nature and, ultimately, of himself. Victor's discovery that a human can be reproduced by means of technology generates an identity crisis. Waldby (2002: 35), in her rereading of *Frankenstein* in the context of nineteenth-century scientific experimentation with vitality, suggests that Victor's fear of the monster indicates that human nature is something always already machinic—that "the human body has been mis-classified as nature all along."

HYPERLINKS AND/AS SCARS

Shelley Jackson's *Patchwork Girl* is in many ways a different tale. Intensely involved with questions of monstrosity, femininity, and reproduction, this is the story of the female mate of Frankenstein's monster, who, having been deleted/aborted by Percy Shelley's editorial pen, is secretly written and sewn together, not by Victor Frankenstein, but by a fictional Mary Shelley. *Patchwork Girl* engages the reader in what neither Mary Shelley nor her male hero could bring themselves to do: reproduce a female monster. The she-monster becomes Mary's lover, travels to America, and, at the age of 175, tapping away on her laptop, has found herself a temporary home in the desert, near Death Valley. *Patchwork Girl* can be read as a simultaneous reproduction and transformation of Shelley's *Frankenstein*, on several levels. *Patchwork Girl* moves the question of what it means to be human from the codex book, with its pages made of paper, to multiple series of interlinked computer-screen images. In this transition, the he-monster becomes a she-monster, and the overall heterosexual framework gives way to oppositional lesbian politics.

Stories written in hypertext are perhaps best thought of as "lexias" (Landow 1997), a term borrowed from Roland Barthes to describe reading units. Whereas a book's paper pages are bound together in a determinate sequence, a story written in hypertext often has more than one point of entry, many internal connections, and no clear ending. It may unfold differently with every reading, according to which potential routes are actualized. In *Patchwork Girl* there are two types of overlapping windows: text windows, containing lexias to be explored (in Storyspace, these windows are called "writing spaces"); and map windows, visual representations of the architecture of the writing spaces and their neighbors. Thus hypertext fiction is inherently intimate with questions of spatiality and mapping. There are no page numbers to refer to in hypertext quotations, only names of writing spaces and the relations between them. The names of the writing spaces are marked throughout the text by <angle brackets>. In a reflexive passage on hypertextual structures, Jackson writes:

> Assembling these patched words in an electronic space, I feel half-blind, as if the entire text is within reach, but because of some myopic condition I am only familiar with from dreams, I can see only that part most immediately before me, and have no sense of how that part relates to the rest. When I open a book I know where I am, which is restful. My reading is spatial and even volumetric. I tell myself, I am a third of the way down through a rectangular solid, I am a quarter of the way down the page, I am here on the page, here on this line, here, here, here. But where am I now? (<this writing>).

Joyce suggests that such self-reflexive passages, typical of hypertext fiction, are something more than postmodern language games: "They do not, I think, merely mark a fledgling stream (a flow is hardly a tradition) of a passing form in an uncertain medium. Instead, or more accurately concurrently, these passages are also a gesture toward a parallel system of reading which invites the reader to read as the writer does rereading" (Joyce 1997: 592). Although the metaphor of linearity is hard to escape, hypertext readings are not readings along the line alone. They carry the possibility of cyclicity, of reading and rereading in other ways, to disrupt and transform the line and sometimes lose it from sight altogether. Apart from those (metaphorical) motions that readings always bring to texts of all kinds, hypertext fiction cannot be read without also being (mechanically) reproduced by the reader. This distinction between the metaphorical and the literal reproduction of texts is at the heart of what Aarseth (1997) calls "cybertext," which he understands as the ways in which the reader "completes" the text, and as the text's partial self-completion.

The point is that it is not the structure of hypertext fiction that makes it unique; rather, a work like *Patchwork Girl* is tied to its medium in a sense that makes the two inseparable. When texts move from written pages to screened performances, it is necessary to find ways to take into account both the change of medium and the change of matter. It is important to include an awareness of how acts of reading are always confronted with the viscosity of mediation—of how every medium leaves marks in the processes of textual production. While Aarseth's cybertext is very sensitive to the material specificity of the medium and to the performative dimension of textuality, his "human operator" remains curiously neutral, detached, and disembodied.

Reading a work like *Patchwork Girl* is a corporeal as much as a mindful endeavor. Interlinkages between and among text, body, and machine are always present in acts of writing/reading, but they become explicitly intimate when texts are digitized. In her quest for media-specific analysis, Hayles, a literary theorist, points out that when print fiction, with its durable inscription, is transformed into electronic text, the text becomes mutable in such a way that it renders itself open to on-screen changes: "Such changes imply that the body represented within the virtual space is always already mutated, joined through a flexible, multilayered interface with the reader's body on the other side of the screen" (Hayles 2000: 30). The body of the female monster created through digital texts and images becomes entwined with the body of the reader through his or her physical engagement with the computer (the interactions between fingertips touching the keyboard, eyes moving across the screen, and the Storyspace software in which *Patchwork Girl* was written and in which it is also read). Thus an information technology comes to act as a reproductive technology, reproducing not only texts and images but also the very life of the she-monster.

The first writing space in *Patchwork Girl* is called <her> and shows a white woman with visible scars all over her body, against a black background crossed by a fractured white diagonal line (see fig. 10.1). This, the reader will learn, is the she-monster before she falls into pieces. Reading, then, becomes an art of sewing and stitching, which reproduces the story as well as the body of the she-monster. Pieces of text as well as pieces of bodies are sewn together, and the heterogeneous origins of these pieces are always visible through the scars and stitches. Scars, in their capacity of simultaneously marking a cut and showing a joining, become the quintessence of the monster's fractured subjectivity: "Scar tissue does more than flaunt its strength by chronicling the assaults it had withstood. Scar tissue is new growth. And it is tougher than skin innocent of the blade" (<cut>).

The scars traversing the monster's body parallel the meaning of the hypertextual links between the writing spaces. But the scars/links do more than make fleet-

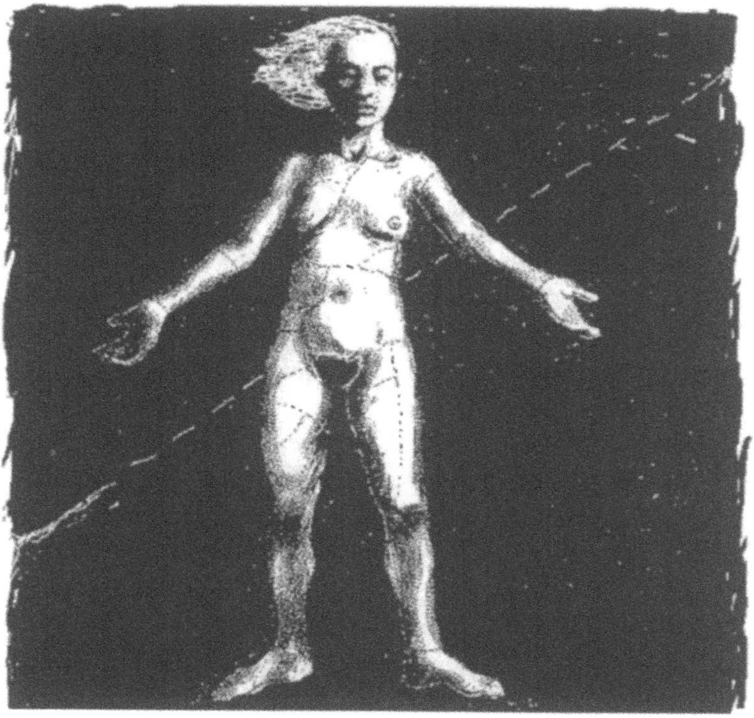

Fig. 10.1. <her>. Image courtesy of Eastgate; see www.eastgate.com

ing overviews possible; they also realize new meanings by putting together and contrasting disparate patches of skin/text. The scars/links simultaneously hold the body/narrative together and pull it apart. They suggest that the whole may be better understood elsewhere, in a different location, or perhaps in the gaps and breaches in between locations. The monster ponders: "My real skeleton is made of scars: a web that traverses me in three dimensions. What holds me together is what marks my dispersal. I am most myself in the gaps between my parts" (<dispersed>).

An understanding of the embodied subject as significantly fractured—as fractured in significant ways—is crucial to an understanding of the importance of links in hypertext fiction (see fig. 10.2). If, as the monster suggests, the reader needs to follow and understand the monster's scars in order to get her, the reader likewise must pay attention to how the links, in their ability to juxtapose—to bring together

Fig. 10.2. The fractured body. Image courtesy of Eastgate; see www.eastgate.com

parts and fragments that otherwise would stay apart—are perhaps more important than the parts themselves. To use links as a way of reading between lexias is to give a hypertexed reading between the lines, as it were.

OF WOMEN BORN

<Her> is linked to <title page>, which reads, "*Patchwork Girl: Or, a Modern Monster*. By Mary/Shelley & Herself," suggesting that the work was co-authored by at least three women: Mary Shelley/Shelley Jackson (both divided and connected), and Herself, supposedly the she-monster. This collaboration and its implied critique of the "origin," or perhaps rather the "originality," of literature works on several levels, letting Mary Shelley figure as writer and creator of the monster, who is

by turns an author, a narrator, and a character. Add to this textual choir the voice(s) of Shelley Jackson, most obviously present in reflexive passages on (hypertext) writing and textual sources.[4] In addition to these three authors, it seems sensible to include a fourth—namely, the hypertext reader. This is not to suggest that the "reader-author" becomes an author on equal terms with the "author-author" but rather that she or he, through acts of hypertext navigation/reading, becomes an author of sorts in causing the work to materialize. The monster ponders, "Mary writes, I write, we write, but who is really writing? Ghost writers are the only kind there are" (<am I mary>). But, as Gilbert and Gubar (1979: 224–25) point out, even Shelley's *Frankenstein* was "a literary jigsaw puzzle, a collection of apparently random documents from whose juxtaposition the scholar-detective must infer a meaning, [consisting] of three 'concentric circles' of narration (Walter's letters, Victor Frankenstein's recital to Walter, and the Monster's speech to Frankenstein), within which are embedded pockets of digression containing other miniature narratives."

It is important to point out that multivocality is not a hypertextual inheritance per se, nor is the idea of the reader as author. *Patchwork Girl* was born when hypertext theory had reached its peak. At the time, many hypertext authors were also hypertext theorists (or the other way around), delightfully mingling postmodern literary theory with hypertext literary practices, to the point where hyperfiction came to be seen as the very epitome of postmodernism. *Patchwork Girl*, as a progeny of this moment in the history of hyperfiction, takes great pleasure in the process of letting the narrative capacities of the medium merge with the narratives being told through writing spaces, the most obvious example being the mirroring of scars and links. For hypertext theorists, postmodern ideas about "the open text," the text as a web of intertextual references, the liberation of the reader from authorial intentions, the active interpretation, and so on, seem to be realized quite literally through the digital text (Bolter 1991; Landow 1997; Landow and Delany 1993; Lanham 1993). Elsewhere (Sundén 2003) I have discussed the problems involved in this confusion of literary theory with digital textuality, problems caused mainly by a curious slippage between (textual) ideology and interface qualities.

SEAMS OF LIFE

The relationship between Mary and the monster is intriguingly multilayered. Like authors falling in love with their own creations, Mary falls in love with her monster—at once her daughter in need of maternal protection, and a fully grown woman. The relation between Mary and her monster is one of deep ambivalence, and there are

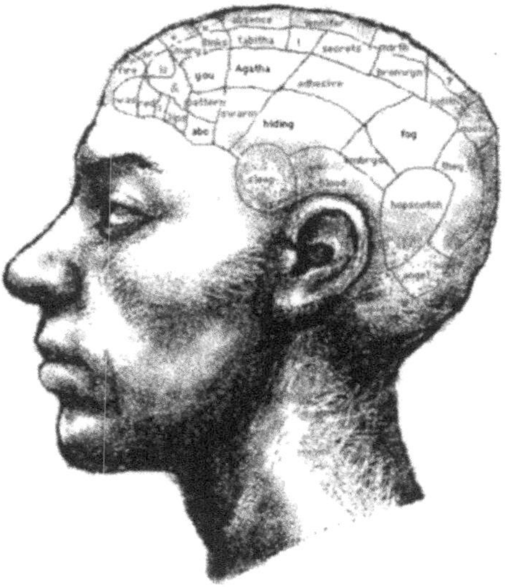

Fig. 10.3. A phrenological drawing of the monster's head. Image courtesy of Eastgate (www.eastgate.com)

strong incestuous indications in the ambiguous bedroom scene (see Ferreira 2000). But these passages, apart from carrying suggestions of incest, are equally strong in their way of conveying images of female lovers. The theme of the she-monster as woman and lover of women cuts across *Patchwork Girl*, making her monstrosity all the more monstrous. She may be viewed as an infant in her initial shyness, but she is also something powerfully different. Her physical strength, and the scars on her body, which still frighten her creator, make her a "daughter" of a different kind, inheriting powers and knowledge that in many ways exceed those of her "mother." In a sense similar to the one in which reading practices in postmodern literary theory are reproductions of the text without the Author, *Patchwork Girl* is deeply engaged with reproduction (of both body and text) without the Father, as it were.

Patchwork Girl has five points of entry—<Graveyard>, <Journal>, <Quilt>, <Story>, and an entrance called <& broken accents>—and all of them make the reader "pass through" the monster's body in various stages of disintegration as her parts rearrange themselves. The last of these entrances is linked to a phrenological drawing of the monster's head (fig. 10.3), which leads to metatextual reflections on writing practices.[5]

Hayles underscores how this passage through the images of the monster's body is necessary in order for the reader to reach the textual writing spaces, a process that reverses the common notion in print fiction of bodies always being represented within the text. In opposition to the idea of fictional bodies as the reader's imaginary fabrications, the body in *Patchwork Girl* "is figured not as the product of the immaterial work but a portal to it, thus inverting the usual hierarchy that puts mind first" (Hayles 2000: 27).

<Graveyard> is where the she-monster is born. The grave becomes a cradle of sorts, where the reader encounters the monster's body in pieces. The writing space <graveyard> reads: "I am buried here. You can resurrect me, but only piecemeal. If you want to see the whole, you will have to sew me together yourself," a statement that leads to <headstone>:

> Here Lies a Head,
> Trunk, Arms (Right
> and Left), and Legs
> (Right and Left)
> as well as divers
> Organs Appropriately
> Disposed
> May They Rest in Piece

Each body part has links to forgotten stories of all the women (plus one man and a cow) who have contributed to the monster's creation. They are the monster's family, uneasily joined, rubbing against each other to create an assemblage of monstrous subjectivity. Like Victor Frankenstein, who learned the secrets of life by way of corporeal decay, the reader of *Patchwork Girl* needs to dig up the bits and pieces of buried body-narratives that together create the monster's origin: "Burdened with body parts, your fingernails packed with mud and chips of bone, you slink out of the graveyard. A kind of resurrection has taken place" (<out>).[6] Reading as needlework quite literally turns bits of death into bits of life.

SUBJECTIVITY AS DECOMPOSITION/RESURRECTION

Patchwork Girl does not fit the idea of transparent, dissolving language that leaves the reader with a seamless illusion of reality. It is all about seams, from the obvious collage techniques in the <quilt> section, where each patch is itself made of several smaller patches, to the metalevel of hypertext montage. The monster's seams show how what are conventionally thought of as coherent and autonomous texts and bodies are in fact pieced together from many different sources, and this fact causes the story to shuttle back and forth between textuality and genetics. Not even our body is our own; rather, it is "haunted by our uncle's nose, our grandfather's cleft palate, our grandmother's poor vision, our father's baldness. There are ghosts in the form of recessive genes, that never show themselves to us, but might appear to our children, to the seventh son of the seventh son. Red hair, suddenly, out of a clear blond lineage" (<body ghosts>).

Patchwork Girl treats language as material, even corporeal, to the point where it is no longer possible, or even meaningful, to distinguish among the reproduction of texts, bodies, and quilts:

> I had made her, writing deep into the night by candlelight, until tiny black letters blurred into stitches and I began to feel that I was sewing a great quilt, as the old women in town do night after night . . . and their strokes grow thicker than machinery and tight enough to score deep creases in the cloth. I have looked with reciprocal coolness their way, not wondering what stories joined the fragments in their workbaskets [<sewn>].

Patchwork Girl is an intermingling of traditional women's arts, such as sewing and quilting, with high-tech, cyborgian, queer performances. The monster struggles endlessly for connectivity and a sense of wholeness, and even though she is never able to leave her position as an outsider, she manages quite easily, by contrast with Frankenstein's monster, to create transitory connections with other women and monsters on the cultural margins. But the harder she tries to be in control of her incongruous parts and subjectivities, to master the die-hard fiction of the Enlightenment era unified, autonomous self, the more her body parts and the ghostly stories of their donors fight back. She starts losing limbs and has great trouble gluing them back on. One day in the bath with Elsie (one of the she-monster's lovers), she comes completely apart and starts learning how to live and love through a body of differences:

> My parts bobbed in changing patterns in a warm reddish slurry of bathwater and blood. Elsie was immersed in me, surrounded by fragments, but somehow she held me. I was gathered together loosely in her attention in a way that was interesting to me, for I was all in pieces, yet not apart. I felt permitted. I began to invent something new: a way to hang

together without pretending I was whole. Something between higgledy-piggledy and eternal sphere [<I made myself over>].

In its constant decomposition of texts and bodies, *Patchwork Girl* can be seen as an efficient deconstruction of the subject—as a powerful metaphor for how bodies are always partial existences, with diverse and unclear origins. Following the fragmented aesthetic of hypertext, the images in *Patchwork Girl* show the body of the female monster as scarred and in pieces. Although *Patchwork Girl* is a postmodern tale of the deconstruction of Enlightenment subjectivity—of the self as unitary, independent, and thoroughly rational—it is simultaneously a clever reconstruction of an alternative embodied female subjectivity (see chapter 12, this volume). The stories of the she-monster are not only about the body/text as deconstructed and torn apart; "she" also has a great deal to say about transformation and creativity, about putting things back together in different ways. Ultimately, "she" is about a process of learning to hang together while still in pieces, of learning to accept fractures and fissures as a way of understanding and moving through a (post)modern world.

In a similar vein, the reader of *Patchwork Girl* is initially invited to "resurrect" the monster by digging up her various body parts through multiple graveyard stories and putting her together. As the reader travels through the landscape of broken texts and images, the she-monster's struggle for a sense of coherence and completeness is likely to be mirrored by the reader's struggle for meaning and narrative wholeness.[7] Where does the story end? Where does it begin? Where am I? What have I missed? What's the point, anyway?

Hypertexed experiences have the potential to pass through a range of existential questions and fears. As Hayles (2000: 1) has it, "Five hundred years of print have made the conventions of the book transparent to us," and she points out how digital modes of literature shed light on the situatedness and material specificity of all writing technologies (ultimately revealing the habitual safety of the printed page as merely illusionist). Reading, then, is never safe. It is precisely this implicit lack of safety that becomes explicit in hypertext stories. Like the she-monster—who finds a form of fleeting acceptance in a self-identity as assemblage, a way of hanging together that is fundamentally a rifted, (dis)continuous becoming—the reader needs to find navigational strategies that explore and create new senses of beginning as well as new senses of ending. And this kind of end may turn out, in the end, to be no end at all but rather a process of finding out: "If you think you're going to follow me, you'll have to learn to move the way I do, think the way I think; there's just no way around it. And then . . . you'll begin to have trouble telling me apart from yourself" (<think me>). As the reader learns to move the way the monster moves, to give up the search for the True Story—which never existed anyway—a shift in the reader's subjectivity is already taking place. In contrast to Victor Frankenstein, "she" has already given in to her own machinic monstrosity.

POSTHUMAN PLEASURE AND PAIN

As tempting as it might be to read *Patchwork Girl* as a postmodern celebration of a poetics of the fragment (see Harter 1996), this work also speaks of experiences of being fragmented as severely painful. *Patchwork Girl*, like *Frankenstein*, is about the limits of what it means to be human. It is about the struggle to come to terms with difficult feelings of being made, not born: "Authenticity was the surest advertisement. Born monstrous, we would have had a kind of wholeness. . . . Manmade we were forgeries. And while patchwork had its period in vogue, patched freaks, conglomerates, never did" (<manmade>). In embodying a profoundly ambivalent balancing act between the pain and the pleasure of fragmentation, *Patchwork Girl* is an unusual posthuman narrative of the ever longed-for but never fulfilled dream of unification, and of feelings of grief and loss related to prosthetically mediated experiences (see Ferreira 2000).

When the she-monster meets Elsie for the first time, she buys her past, in a vain attempt to secure a coherent personal narrative. Like the android Rachel in Ridley Scott's 1982 film *Blade Runner*, she knows that she needs visual memories in order to be "real": "Her past was perfect for me. It had little black corners and layers of snapshots sliding loose on top of the earlier grid" (<photo album>). As Doane (2000) points out, reproduction is a guarantee of history (and, one might add, of the future). Human reproduction organizes generations, whereas mechanical reproduction structures memories. For Rachel images of her mother in particular are what will guarantee her "realness." Technologies of reproduction work, in a sense, to control the excesses of the maternal. When this control is exercised—whether as birth control, in vitro fertilization, or the Frankensteinian "death" of the mother—reproductive technologies not only make possible new parental definitions and practices but also weaken the more positive connotations of the maternal: "Without her, the story of origins vacillates, narrative vacillates. It is as though the association with a body were the only way to stabilize reproduction" (Doane 2000: 120).

The question is, What happens to this epistemological certainty—the fact that the mother is knowable, whereas the father is uncertain until genetically validated—when what it means to be a mother also starts to vacillate, not to say replicate? If association with a body is the only way of stabilizing reproductive anxiety, is such stabilization possible when "the mother" can be many? As with the multiple maternal functions in *Patchwork Girl* (featuring Mary Shelley and Shelley Jackson along with all the female body/narrative donors in addition to one man and a xenotransplant from a cow), motherhood may be shared among, for example, egg donors, surrogate mothers and "birth" mothers, and "social" mothers. If, at some level, the fact of having a mother is used to guarantee certainty of historical knowledge—a narrative point of departure—who counts as the privileged mother-person when there are several? Where does the story begin? The answers to these questions may be many and varied, but the answers provided here are all tied to a virtually fatherless, multiply mothered, artificially reproduced she-monster and her striving to find comfort through untraditional narratives:

"My birth takes place more than once. In the plea of a bygone monster; from a muddy hole by corpse-light; under the needle, and under the pen. Or it took place not at all. But if I hope to tell a good story, I must leapfrog out of the muddle of my several births to the day I parted for the last time with the author of my being, and set out to write my own destiny" (<birth>).

NOTES

1. See also the argument in Sundén (2003) for a "she-cyborg"—a productive feminist (sexually "differentiated") coupling between women and machines. This chapter can be read as a continuation of that discussion, with the girl monster of Shelley Jackson's *Patchwork Girl* (1995) posing as a she-cyborg.

2. Homans (1986: 113) takes the argument for Frankenstein as a horror story of maternity one step further by reading the novel as a "collision between androcentric and gynocentric theories of creation, a collision that results in the denigration of maternal childbearing through its circumvention by male creation." She draws attention to the fact that the monster is not a woman-born child but a man-made creation, yet she remains sensitive to how the masculine desire to be in control of one's (technological) creations is profoundly disturbed by a technology that gets out of control. Compare Huyssen (1986: 70), who argues, "The ultimate technological fantasy is creation without the mother."

3. There is a rarely acknowledged "materialistic" side to Mary Shelley's feminism. As Youngquist (1991) points out, many of Shelley's feminist critics are the intellectual heirs of Shelley's mother, the liberal feminist Mary Wollestonecraft, and so they fail to take into account the ways in which *Frankenstein* can be read as a critique of liberal feminism. The meanings and the matter of the monster's

corporeality, completely at odds with the Enlightenment emphasis on disembodied reason, are at the heart of the story. Although Shelley's creation of the monster borders on a determinist notion of the body as fate, Youngquist shows how the monster is both the most reasonable character in *Frankenstein* and the character most "cursed by the brute fact of bodily existence. . . . The monster demonstrates openly the imperatives of corporeal life: there can be no transcendence of sex, no rationalist utopia oblivious to the body" (Youngquist 1991: 342–44).

4. See Leader (1996: 167–205) for a discussion of collaborative (in this case, dual) authorship in the "parenting" of *Frankenstein*. Then again, that "collaboration" was of a very different kind than the suggested co-construction of *Patchwork Girl*, consisting merely of Percy Shelley's extensive revisions of Mary Shelley's manuscript.

5. <Graveyard> is the monster's birthplace. <Journal> is Mary Shelley's journal about her monster. <Quilt> is a cutting apart and pasting together of quotes in crazy-quilt style, including pieces of Shelley's *Frankenstein*, L. Frank Baum's *Patchwork Girl of Oz*, *Elle* magazine, the software manual *Getting Started with Storyspace*, and scraps of Donna Haraway, Hélène Cixous, Jacques Derrida, and others. <Story> is the monster's story.

6. For a parallel reading of Shelley's *Frankenstein* and the science of anatomy in the legal context of early-nineteenth-century grave robbing ("resurrection"), see Marshall (1995).

7. Hypertext criticism, like literary criticism in general, has been preoccupied with texts and (scholarly) readings. While "close readings" of hyperworks are essential, surprisingly few critics have been equally close to readers, to their situated hypertext experiences and ways of engaging with the work. For attempts in this direction, see Douglas (2001), Moulthrop (1991), Moulthrop and Kaplan (1991), and Sundén (2006).

PART 4

Philosophies of Life

11 ▪ Living in a Posthumanist Material World

Lessons from Schrödinger's Cat

KAREN BARAD

In preparing the material that eventually became this chapter, I went in search of a little book called *What Is Life?* (Schrödinger [1944] 1967), authored by one of the founders of quantum physics. A nagging sensation in my brain kept telling me I needed to find the book. I tried to shake off the feeling, but it wouldn't release me. I was caught between a desire to get my hands on this book and the visceral sense that it already held me in its grip. I went into my local bookstore to inquire about the title and cheerfully spelled "Schrödinger" for the clerk. Without looking up from his computer terminal, he replied that he didn't have it, but then, with sudden animated attention, he lifted his face toward me and added, "Is that the guy with the weird cat?" "Yes," I replied heading out the door, "he's the guy with the weird cat."

Schrödinger's *What Is Life?* has been described as one of the most influential writings of the twentieth century, and some regard it as the defining contribution of an Oedipal birthing of the new biology by modern physics. Evelyn Fox Keller (1992) and other historians of science have remarked on the extraordinary effect that physicists had on the development of the field of molecular biology:

> It begins with the claim of a few physicists—most notably, Erwin Schroedinger [Schrödinger], Max Delbruck, and Leo Szilard—that the time was ripe to extend the promise of physics for clear and precise knowledge to the last frontier: the problem of life. . . . Emboldened by the example of these physicists, two especially brave young scientific adventurers, James Watson and Francis Crick, took up the challenge and did in fact succeed in a feat that could be described as vanquishing nature's ultimate and definitive stronghold. Just twenty years earlier Niels Bohr had argued that one of the principal lessons taught by quantum mechanics was that "the minimal freedom we must allow the organism will be just large enough to permit it, so to say, to hide its ultimate secrets from us" (Bohr 1958).

> Now, as if in direct refutation of Bohr's more circumspect suggestion, Watson and Crick showed, with the discovery of DNA, and accordingly, of the mechanism of genetic replication, that areas apparently too mysterious to be explained by physics and chemistry could in fact be so explained . . . [and so] Watson and Crick embarked on a quest that they themselves described as a "calculated assault on the secret of life" [Keller 1992: 41–42].

In *Secrets of Life, Secrets of Death*, Keller (1992) examines the parallel functioning of secrets in matters related to the discovery of the structure of DNA and the making of the atomic bomb. She argues that perhaps the most important effect that physicists had on the development of contemporary biology was not the transfer of any particular technology—instrumental or mathematical—from physics to biology so much as a transformation of the biologists' philosophical stance toward the question of life: "To historians of science, the story of real interest might be said to lie in the redefinition of what a scientific biology meant; in the story of the transformation of biology from a science in which the language of mystery had a place not only legitimate but functional, to a different kind of science—a science more like physics, predicated on the conviction that the mysteries of life were there to be unraveled, a science that tolerated no secrets" (Keller 1992: 42).

Secrets, in their flirtatious play of the visible and the invisible, reveal something of the erotics of knowledge seeking and the seductive lure of epistemological temptation. But the persistence of this metaphor, the hold it has on our technoscientific imaginary, may ultimately reveal as much about ethics as about epistemology.[1] From the *Enola Gay* (the aircraft that hid the Manhattan Project's "baby" in its "womb" before revealing it to the world in an explosive announcement that shattered the lives of tens of thousands living in Hiroshima and Nagasaki[2]) to ultrasound technologies (touted by antiabortion groups as disclosing the hidden Truth of Life, laid bare to the masculine gaze of science), the image of science as the great revealer persists, despite its expanding role as the great technician; mastery of Nature's secrets and her re-creation in Man's image go hand in hand. But what makes us think there's any secret to be told—a secret to be revealed by science, or to remain concealed in a shroud of mystery? What ontological presuppositions serve as a foundation for such thinking? What kinds of geometrical relations of depth and surface, inside and outside, thinking and being, are presumed? What accounts for the persistent epistemological screenplay that casts Nature as the coy female awaiting the heroic advance of the masculine scientist who will lift her veil and expose her secrets? What kinds of ethical stances are fostered by or even founded on such epistemological and ontological assumptions? What makes us think that Nature keeps secrets? What if there were no secret to tell, no hidden states of affairs to reveal? How would we understand the role of science and technology? Would science lose its purpose, its edge, its raison d'être?

There is an entangled tale to tell about the connections between and among bombs, babies, DNA, quantum phenomena, physicists, and biologists. I'm not going to add anything here to the illuminating analyses of these entanglements that have been offered by Keller and others. Rather, what I propose to do in this chapter is to say a few words about the nature of entanglements themselves, and about how an investigation of their nature points to an undoing of the very foundation of secrecy. I begin with that ever so weird cat of Schrödinger's.

CAT PARADOX

> A cat is penned up in a steel chamber [see fig. 11.1], along with the following diabolical device (which must be secured against direct interference by the cat): in a Geiger counter there is a tiny bit of radioactive substance, so small that perhaps in the course of one hour one of the atoms decays, but also, with equal probability, [it does not decay]. [If it happens that an atom decays], the [Geiger] counter tube discharges and through a relay releases a hammer which shatters a small flask of hydrocyanic acid. If one has left this entire system to itself for an hour, one would say that the cat still lives if meanwhile no atom has decayed. The first atomic decay would have poisoned it. The ψ-function of the entire system would express this by having in it the living and the dead cat (pardon the expression) mixed or smeared out in equal parts [Schrödinger (1935) 1983].

Let's consider Schrödinger's description of his famous cat-in-the-box thought experiment. The notion of probability is key in this experiment. In order to appreciate the role it plays, let's begin by reviewing some key features of classical or Newtonian physics.

According to Newtonian physics, nature is both determinate and deterministic: objects exist with determinate boundaries and properties, and the entire future (past) of the objects can be predicted (retrodicted) with certainty on the basis of the measured values of pre-existing states of the objects at a given moment in time (for example, an object's position and momentum), as obtained through measurements that reveal these pre-existing values. The correlative worldview holds that objective knowledge reveals the objective state of the world. Physicists investigating events on an atomic scale in the first part of the twentieth century saw many

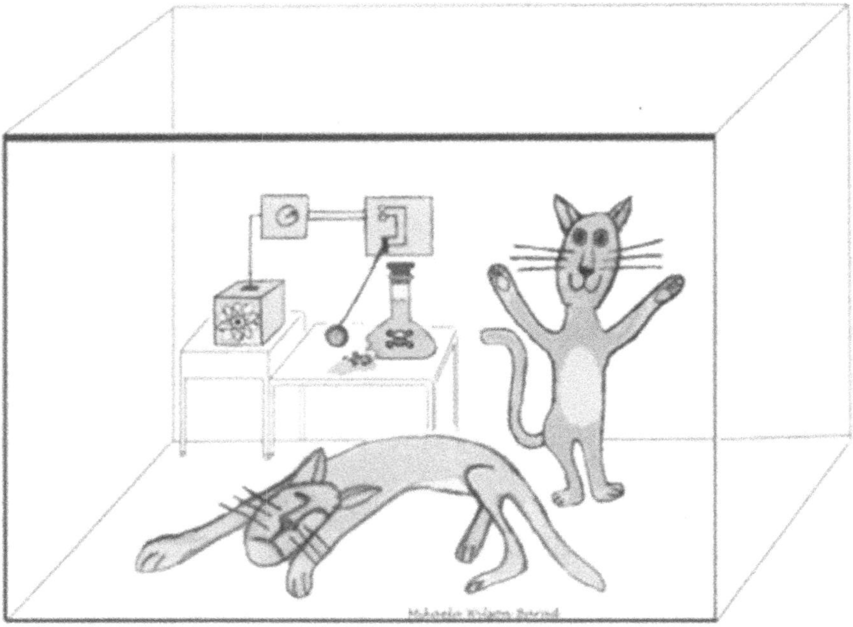

Fig. 11.1. Schematic illustration of Schrödinger's cat experiment. Illustration commissioned by the author and created by Mikaela Wilson-Barad; used with permission.

difficulties with classical Newtonian physics and proposed a successor theory, quantum physics, which seems to suggest that nature does not exist in fully determinate states, and that what can be predicted is only the probabilities for particular events to occur. That is, quantum theory seems to reject both the determinate nature of existence and strict causal determinism.[3]

Let's turn our attention now to the cat experiment. According to quantum theory, it is not possible to predict with *certainty* whether any given atom will decay in a given period of time, but it is possible to use the Schrödinger equation to calculate the *probability* that an atom will decay in a given time period. In the specific case of Schrödinger's cat experiment, a radioactive substance is used that has a 50 percent probability of decaying in one hour. Furthermore, the system is set up in such a way that the cat's fate is coupled to the fate of the atom; that is, their two fates are *entangled*: if the atom decays, the cat is killed, and if the atom doesn't decay, the cat remains alive. Schrödinger's equation predicts that the overall state of the entangled system after one hour is a superposition of two states—a nondecayed atom together with a live cat, and a decayed atom together with a dead cat—with either possibility being equally probable.

You may be wondering what, exactly, a superposition or entangled state is. And so do physicists. The cat paradox is Schrödinger's rhetorical device for issuing a wake-up call to his fellow physicists, who were, in his opinion, all too complacent about the fact that electrons and other microentities are commonly in such states of superposition. Schrödinger's hope was that if he dramatized this situation by replacing the common exemplar of a microscopic particle with a cat, he could call attention to the peculiar nature of entangled states. What can it mean for a cat to be in a superimposed state of dead and alive?

I will address this question in a moment, but first I want to point out that this is just the tip of the iceberg, the first glimpse into a Pandora's box of quantum quandaries, both epistemological and ontological, that the cat-in-the-box and other such thought experiments raise. The central quandary of the cat experiment is this: if we look inside the box after one hour, we won't find the cat in a superimposed state of dead *and* alive; rather, it will be *either* dead *or* alive. But this is not what the Schrödinger equation predicts! The Schrödinger equation leaves us with the superposition and doesn't account for its resolution into one definite state or another (that is, dead or alive) upon measurement. This fact has spawned innumerable creative interpretations of quantum physics, including the shockingly anthropocentric hypothesis that human consciousness collapses the superposition into one definite state or another, and the bizarre, metaphysically hyperschizophrenic "many worlds" interpretation in which each measurement that is performed splits the world into multiple parallel universes that are realizations of each possibility and are entirely inaccessible to one another (in which case the cat in question is alive in some worlds and not in others). Barad (2007) explains the nature of entanglements and proposes a resolution of the cat paradox. For our purposes here, it is sufficient to summarize a few key points.[4]

First of all, given the multitude of misconceptions and misstatements about the cat paradox, I want to begin by dispelling some incorrect understandings of this "smearing" or "blurring" of the cat. Despite what is presented in some popular accounts, it is not the case (1) that the cat is *either* alive *or* dead (we simply do not know which), or (2) that the cat is *both* alive *and* dead simultaneously (this possibility is logically excluded, since "alive" and "dead" are taken to be mutually exclusive states), or (3) that the cat is *partly* alive and *partly* dead (a kitty in a coma), or

(4) that the cat is in a state of being *neither* alive *nor* dead (perhaps in the sense of a vampire cat living among other "undead" creatures).[5] Rather, I argue that the correct way to understand what this superposition (or "blurring") stands for is to understand that *the cat's fate is entangled with the radioactive source*—and not merely epistemically, as Schrödinger and others suggest, but *ontically*; that is, the cat and the atom do not *have* separately determinate states of *existence*, and, indeed, there is no determinately bounded and propertied entity that we normally identify with the word "cat," independently of some measurement that resolves the indeterminacy and specifies the appropriate referents for the concepts of "cat" and "life state." That is, contrary to our classical sensibilities, which tell us that cats and other animate entities are determinately bounded entities that are either dead or alive, independently of any measurement of their "life state," the fact is that *no* fact of the matter concerning the life state of the cat exists independently of some way of measuring/defining "life state" (which resolves the indeterminacy of the boundaries and properties in question).[6] That is, it is *not* the case that the cat is simultaneously dead and alive; "it" simply has no determinate life state—that is, there is no determinate fact of the matter about whether "it" is dead or alive. Indeed, in the absence of necessary defining conditions, the very notion of a "life state" is not well defined—it is without any determinate meaning.

As I have argued elsewhere (Barad 2003, 2007), quantum mechanics deconstructs the metaphysical substrate of representationalism—the belief that words and things are primary units with inherently determinate boundaries, properties, and meanings. Indeed, in recent years, quantum physics has yielded compelling empirical evidence for the surprising claim that things do not have inherently determinate boundaries and properties, and that words do not have inherently determinate meanings. According to agential realism, we err in helping ourselves to the notion of a "life state" (state of "aliveness") and to the presumption of its inherently determinate nature. By my agential realist account, concepts like "life state" or "aliveness" are not merely ideational; rather, they are specific material configurations. And the semantic and ontological indeterminacy is resolvable only through the existence of a specific material arrangement that gives meaning to particular concepts to the exclusion of others, thereby effecting a cut between "object" and "subject," neither of which pre-exists their intra-action.[7] The notion of intra-action (as opposed to the more usual "interaction," which presumes the prior existence of independently determinate entities) entails a profound shift in the epistemological and ontological landscape, including changes in the nature of causality and agency.[8]

What does it mean to suggest that the notion of "life" or "aliveness" must be materially specified in order for the "cat" to be dead or alive? Is there not some natural, obvious, or inevitable definition of "life" that we might presume? Carl Sagan, writing the entry on "life" for the 1973 edition of the *Encyclopedia Brittanica*, drew attention to five different definitions of life, based on physiology, metabolism, biochemistry, genetics, and thermodynamics.[9] And that was simply for the carbon-based organisms that biologists like to fancy as the referent point for such discussions. Of course, 1973 was a time when it was widely accepted that the very question of the nature of life was one that could be answered by biologists. Now, not only has our taxonomy of living entities grown, but the usual "dualisms between structure and process, form and function, part and whole, inheritance and environment, contingency and necessity, holism and reductionism, vitalism and mechanism, energy and information, concept and metaphor" (Emmeche 1994: 560) are no longer sustainable. Clearly, questions about how to define life differ for a carbon-based organism like a cat; a human fetus; a genetically modified translife form that crosses the animal and plant kingdoms, such as a "fishberry"; a silicon-based, artificial intelligence–programmed being like Kismit; an artificial life form like a snail "animat"; or unknown life forms from other planets (see fig. 11.2).[10] Let's face it—life just ain't what it used to be, if it ever was.

And yet the question of how to define life is not the only crucial issue for these considerations. Suppose we make a device that measures and defines "life" or "aliveness" according to one definition or another, and that we use it to detect the fate of whatever it is that's locked up in the box with the radioactive source. Once we've performed such a measurement, we are left with the question of what it is that we've measured.

I've already noted that measurements do not reveal the already determinate state of a pre-existing entity. Rather, it is through specific measurement intra-actions that the boundaries and properties of the "measured object" and "measuring instrument" become determinate, and that particular embodied concepts (like "life") become meaningful. So what is the objective referent for the measured property? The referent is not some measurement-independent, determinately bounded object (which doesn't exist) but rather the *phenomenon*—the entanglement of intra-acting agencies ("objects" and "subjects"). This referent includes much more than we could ever draw in any diagram; in particular, it includes the entangled and enfolded sets of apparatuses of bodily production of all the beings and devices relevant to this specific example.[11]

By my agential realist account, human consciousness is not the cause of the pro-

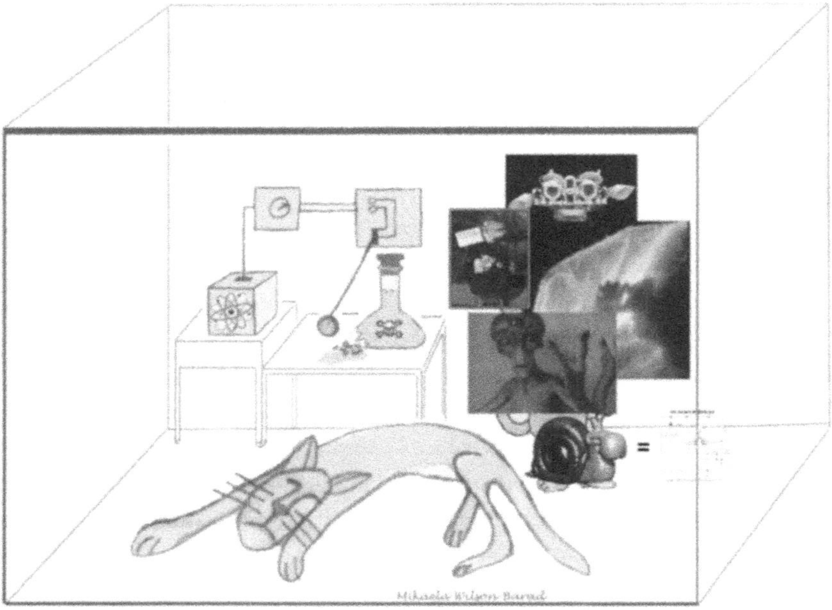

Fig. 11.2. Schematic illustration of Schrödinger's cat experiment, modified to suggest the inherent indeterminacy of the notion of "life" or "aliveness." How is "aliveness" to be measured? Will it differ for different life forms? What material-discursive apparatuses constitute a meaningful notion of "aliveness" for a cat? Is it the same for a human fetus? a fishberry? a robot? an animat (creatures featured in studies of Artificial Life)? an alien?

duction of determinate states of being; rather, "humans" must also be understood as phenomena, produced through the intra-action of multiple material-discursive apparatuses of bodily production, and "consciousness" cannot be presumed to be an inherent property of individuals. I can't provide a full explanation here, but elsewhere (Barad 2003) I develop a posthumanist performative account of the dynamics of mattering that goes beyond both humanist and antihumanist accounts of the role of the human. By "posthumanist," I do not mean postmodernist celebrations or demonizations of the "posthuman" or technohuman as testimonies to the death of Man, nor do I mean the next stage of Man. Rather, I use the terms "posthumanist" and "posthumanism" to mark a commitment to accounting for the boundary practices through which the "human" and its others are differentially constituted.

Posthumanist performativity is not a celebration of difference for difference's sake; rather, it is about accountability to and for differences that matter.

By my posthumanist performative account, matter is not a fixed essence or property of things; rather, matter(ing) is a dynamic process of iterative intra-activity. That is, matter is substance in its intra-active becoming—not a thing but a doing, a congealing of agency. The term "matter" refers to the materiality/materialization of *phenomena*, not to an inherent fixed property of independently existing objects.

By my posthumanist account, phenomena are not the mere result of laboratory exercises engineered by human subjects, nor can the apparatuses that produce phenomena be understood as mere observational devices or laboratory instruments. Rather, apparatuses are boundary-drawing practices through which differential boundaries and properties come to matter. Apparatuses are not mere static arrangements in the world; rather, apparatuses are material-discursive configurations of the world, or, more precisely, dynamic (re)configurings of the world—that is, specific agential practices/intra-actions/performances through which specific boundaries are enacted. And phenomena are not mere laboratory constructions; rather, phenomena are differential patterns of mattering produced through complex agential intra-actions of multiple apparatuses of bodily production.

Materiality is a discursive performance of the world, but discursive practices are not reducible to human-based actions. Matter does not serve as a mere support for discourse, nor is it merely the end product of human-based citational practices. Rather, discursive practices are specific material configurations/(re)configurings of the world through which local determinations of boundaries, properties, and meanings are differentially enacted. That is, discursive practices are ongoing agential intra-actions of the world. Meaning is not a property of individual words or groups of words but an ongoing performance of the world in its differential intelligibility. In its causal intra-activity, "part" of the world becomes determinately bounded and propertied in its emergent intelligibility to another "part" of the world.

That is, discursive practices as boundary-making practices are fully implicated in the dynamics of intra-activity through which phenomena come to matter. The dynamics of intra-activity entails matter as an active "agent" in its ongoing materialization. Materiality is discursive (that is, material phenomena are inseparable from the apparatuses of bodily production—matter emerges out of and includes as part of its being the ongoing reconfiguring of boundaries), just as discursive practices are always already material (that is, they are ongoing material [re]configurings of the world). Discursive practices and material phenomena do not stand in a relationship of externality to one another; rather, the material and the discursive are mutually implicated in the dynamics of intra-activity.

Matter is agentive. Perhaps we might say it is "alive" with possibilities. Electrons, dust particles, and rocks, as much as cats, brittle stars, or humans, are complex phenomena—lively configurations of changing possibilities.[12]

This brings us back to the question of the nature of life. What is life? If quantum physics has any insights to offer into the question of life, these are not concerned with revealing Nature's secrets or helping biologists nail down the answer to what life is, once and for all. Science is not a method for pulling back the curtain on Nature and exposing her essence. There is no pre-existing Nature, and we are not Nature's keeper—a steward, as it were, who stands in a relationship of externality to our charge. Humans in their differential becoming are neither outside observers of the world nor the telos or foundation for life itself. "We humans" are not the external other to nature but rather are part of the lifeblood of the universe in its ongoing re-creation, and we must indeed be accountable to and for the lively relationalities of becoming of which we are a part. For far too long, "we" have been invested in epistemologies, ontologies, and theories of ethics that are based on a belief that the universe harbors secrets from us, that "she" is something other than us, something that can't be fully trusted. The word "secret" comes from Latin *secretus*, the past participle of *secernere* (to be separate or set apart). Life is not a secret to be revealed. Life, in all its specific material configurations, is not an inherent property of separate individual entities but rather an entangled agential performance of the world.[13] There is no final answer to the question of life; indeed, there is no final state of aliveness that can be pinned down—which, indeed, is a great tribute to the inexhaustible creative vitality of the world.

NOTES

This chapter is based on a talk titled "What Is Life?" It was prepared for the symposium "What's Life Got to Do with It?" that was held at Lancaster University, June 11, 2004.

1. On an agential realist ethics, see Barad (2007), which argues that epistemological, ontological, and ethical considerations are not separable.

2. Keller (1992: 44) cites Brian Easlea's documentation that "the metaphor of pregnancy and birth became the prevailing metaphor surrounding the production and the testing . . . of the atomic bomb."

3. In the literature, there has been an unfortunate tendency to conflate issues of determinacy and determinism. But the question of the existence of determinate states of nature is not the same as the question

of a deterministic sense of causality. Determinism is the proposition that events follow an unbroken causal chain. In Newtonian physics, the future (and past) state of any system is given by the deterministic laws of nature, which allow the prediction (and retrodiction) of events on the basis of "initial conditions" (the values of the position and the momentum of an object at any given moment of time). Thus, for example, according to Newtonian physics, your reading of this sentence right now was already determined at the moment of the Big Bang (the atoms that make up your body were just following their preset trajectories), whereas the issue of determinacy (and indeterminacy) has to do with the nature of existence—in particular, with whether entities possess definite boundaries and properties (independently of the circumstances of their measurement). In the case of quantum physics, the issues of determinacy and determinism are linked (but nevertheless conceptually nonidentical) in the following way: since the initial conditions cannot be specified, because there is no determinate initial state of affairs (that is, there is no determinate fact of the matter about the values of both the position and the momentum of an object at the same moment of time), the future course of an object cannot be said to evolve in a deterministic fashion from a given moment in time (that is, there is no determinate trajectory of the object). I should perhaps also note that not all interpretations of quantum physics take indeterminacy to be an ontological issue. Some interpret "indeterminacy" (sic) in an epistemological fashion, as a matter of the *uncertainty* of our knowledge (of a presumed determinate state of affairs).

4. One may wonder what to make of the fact that, when it comes to quantum physics, physicists seem willing to place the human at the center of a theory of the universe, despite the fact that such placement seems to fly directly in the face of a more general philosophical stance among physicists, which positions the human as the outside observer of the universe. For a detailed discussion, see Barad (2007); see also Keller (1985).

5. Most often, incorrect accounts of the paradox offer the "simultaneously dead and alive" interpretation and ask us to be shocked and horrified by such an outcome, which, despite its counterintuitive nature (and our own averse reactions), it is alleged to show the bizarre nature of quantum phenomena. But not only does such an account misconstrue the state that the cat is in while in the chamber, it completely misses the crux of the paradox.

6. By contrast, Schrödinger interprets *entanglements* epistemically rather than ontically: what an entangled state represents for Schrödinger is an uncertainty in the state of our knowledge of the system, not an ontological indeterminacy in the state of the system. For further details, see Barad (2007, chap. 7).

7. What is at issue in my agential realist interpretation is not the necessity of measurement per se, at least not in the narrow sense of human-based actions on the world, but rather the necessity of specific material configurations and practices that need not involve, let alone be based on, "humans" (which are also not inherently bounded and propertied entities).

8. The shift from the usual "interaction" to the agential realist "intra-action" also signals a shift in the notions of space, time, matter, discourse, mind, body, nature, culture, thought, action, fact, value, and many others; see Barad (2007).

9. Noted in Emmeche (1994).

10. See, for example, Barad (1998) for an agential realist analysis of some of the

material-discursive apparatuses of bodily production that constitute contemporary conceptions of fetal life. See also Haraway (1997), Hartouni (1997), and many other analyses.

11. Specific examples are given in Barad (2007), where there is also a detailed explication of the notion of an apparatus.

12. The agential realist notion of a phenomenon entails a radical break from the metaphysics of individualism. The world is not made up of individuals with independently determinate properties; rather, phenomena are ontologically primary relations, that is, entanglements of agentially intra-acting lively (re)configurings of the world. While some may argue that the notion of life ought to be reserved for creatures that have the ability to remember, a close examination of creatures without brains, such as brittle stars, calls into question this narrow restriction on the nature of life, mind, and memory. Memory and re-member-ance, I argue, are not purely ideational (mental) capacities but marked/material historialities engrained in the body's becoming (where bodies cannot be presumed to end at the skin). For further discussion, see Barad (2007).

13. One might be inclined to say that life is distributed rather than localized. Nevertheless, while the term "distributed" unsettles the more usual sense of life as a property of individuals, it also has the unfortunate connotation of some fixed quantity that it is dispersed over multiple actors. I want to call into question not only the notions that actors pre-exist, and that agency is a property that defines what it means to be an actor (that is, the notion that agency is something an actor has), but also the idea that life is a fixed or continuous quantity (which is distributed over actors).

12 ▪ The Politics of Life as Bios/Zoe

ROSI BRAIDOTTI

We are witnessing today a proliferation of discourses that take life as a subject and not as the object of social and discursive practices. Discussion of biopolitics and biopower can be considered central to cultural studies, feminist theory, and science and technology studies. In this speculative chapter, I propose the primacy of life as zoe. I oppose zoe, as vitalistic, prehuman, generative life, to bios, as a discursive and political discourse about life. I want to defend the argument that the emergence of these discursive "bits of life" results in the need for more social and intellectual creativity in the scientific as well as the mainstream culture.[1]

THE EMERGENCE OF ZOE

Life is half animal, or zoe (zoology, zoophilic, zoo), and half discursive, or bios (biology). Zoe, of course, is the poor half of a couple that foregrounds bios, defined as intelligent life. Centuries of Christian indoctrination have left a deep mark here. The relationship to animal life, to zoe rather than bios, constitutes one of those qualitative distinctions upon which Western reason erected its empire. Bios is almost holy, whereas zoe is certainly gritty. That they intersect in the human body turns the physical self into a contested space, and into a political arena. Historically, mind-body dualism has functioned as a shortcut through the complexities of this in-between, contested zone. One of the most persistent and helpful fictions told about human life is that of its alleged self-evidence, its implicit worth. Zoe is always second best, and persistence of life independently of rational control, even regardless of it and at times in spite of it, is a dubious privilege attributed to nonhumans, a category that includes all the animal kingdoms as well as the classical "others" of metaphysically based visions of the subject—namely, the sexual other (woman) and the ethnic other (the "native"). In the old regime, this category used to be called "Nature."

Traditionally, self-reflexive control over life is reserved for humans, whereas the mere unfolding of biological sequences is for nonhumans. Given that the concept of "the human" was colonized by phallogocentrism, it has come to be identified with male, white, heterosexual, Christian, property-owning, standard-language-speaking citizens. Zoe marks the outside of this vision of the subject, in spite of the efforts of evolutionary theory to strike a new relationship with the nonhuman. Contemporary scientific practices have forced us to touch the bottom of an inhumanity that connects with the human, and that does so precisely in the immanence of the human's bodily materialism. With the genetic revolution, we can speak of a generalized "becoming infrahuman" on the part of bios. The category of "bios" has cracked under the strain and splintered into a web of interconnected "bits of life" effects (Rose 2001) .

With the postmodern collapse of the qualitative divide between the human and "his" others (the gender is no coincidence), the deep vitality of the embodied self has resurfaced from under the crust of the old metaphysical vision of the subject. Zoe—this obscenity, this life in me—is intrinsic to my being and yet is so much "itself" that it is independent of the will, of the demands and expectations of sovereign consciousness. This zoe makes me tick and yet escapes the control of the supervision of the subject. Zoe carries on relentlessly and gets cast out of the holy precinct of the "I" that demands control and fails to obtain it. Thus zoe ends up being experienced as an alien other. Life is experienced as inhuman because it is all too human, as obscene because it lives on mindlessly. Are we not baffled by this scandal, this wonder, this zoe—that is to say, by an idea of life that exuberantly exceeds bios and supremely ignores logos? Are we not in awe of this piece of flesh called our "body," of this aching meat called our "self," expressing the abject and simultaneously divine potency of life?

Classical philosophy is resolutely on the side of a dialogue with the bio-logical. Nomadic subjectivity, by contrast, is in love with zoe.[2] It's about the posthuman as becoming animal, becoming other, becoming insect—trespassing all metaphysical boundaries. Ultimately, it leads to becoming imperceptible and to fading, with death as just another sequence in time. Therefore, some of these "bits of life" effects are very closely related to that aspect of life which, though it goes by the name of death, is nevertheless an integral part of the bios/zoe process. The bios/zoe compound refers to what was previously known as "life" by introducing a differentiation internal to that category. This differentiation, by making the notion of life more complex, implies the notion of multiplicity. In turn, multiplicity allows for a nonbinary way of positing the relationship between same and other, between different categories of living beings, and, ultimately, between life and death. The emphasis, and hence the mark of "difference," now falls on the "other" of the living body according to its humanistic definition: Thanatos—the death drive, the corpse or spectral other.

This reappraisal of death also means that nowadays the political representation of embodied subjects can no longer be understood within the visual economy of biopolitics, in Foucault's (1976) sense of the term. That is, the (political) representation of embodied subjects is no longer visual in the sense of being scopic, as in the post-Platonic sense of the simulacrum, nor is it specular, as in the psychoanalytic mode of redefining vision within a dialectical scheme of oppositional recognition of self and/as other. Rather, the representation of embodied subjects has become spectral: the body is represented as a self-replicating system that is caught in a visual economy of endless circulation. The contemporary social imaginary is immersed in this logic of boundless circulation and thus is suspended somewhere beyond the life-and-death cycle of the imaged self. Consequently, the social imaginary, led by genetics, has become forensic in its quest for traces of a life that it no longer controls. Contemporary embodied subjects have to be accounted for in terms of their surplus value as genetic containers, on the one hand, and as visual commodities circulating in a global circuit of cash flow, on the other hand. Much of this information is not knowledge-driven but rather media-inflated and thus indistinguishable from sheer entertainment. Today's capital is spectral, and our gaze is forensic.

Accordingly, the human technobodies of postindustrial societies are embedded in complex fields of information, which engender both their explosion into sets of regulatory social practices (dieting, medical control, and pharmaceutical interventions) as well as their implosion as the fetishized and obsessive object of individual concern and care (self-management or all-out prevention of anything that moves). The body is like a sensor, a messenger carrying thousands of communication systems: cardiovascular, respiratory, visual, acoustic, tactile, olfactory, hormonal, psychic, emotional, erotic, and so forth. Coordinated by an inimitable circuit of information transmission, the body is a living recording system, capable of storing and then retrieving the necessary information and of processing it at such speed that it seems to react "instinctively." Fundamentally prone to pleasure, the embodied subject tends toward the recollection and repetition of experiences that pleasure has "fixed," psychically and sensually, on the subject. To remember, after all, is to repeat, and repetition tends to favor that which gave joy and to avoid that which gave pain. The body, as an enfleshed kind of memory, is not only multifunctional but also in some ways multilingual: it speaks through temperature, motion, speed, emotions, and excitement that affects the cardiac rhythm and the like—a living piece of meat activated by electric waves of desire, a script written by the unfolding of genetic encoding, a text composed by the enfolding of external prompts.

LIFE AND THANATOS

I am aware here of forcing the rather unfamiliar opposition of life and Thanatos (historically, of course, the coupled opposites are either "life and death" or "eros and Thanatos").[3] At any rate, in mainstream philosophy there are many different variations being played on the theme of the explosion of bios/zoe. In the work of Giorgio Agamben, for instance, zoe is readily assimilated to the economy of non-life, in the nonhuman sense of the term, be it in the animal, the vegetable, or the machinic sense. More specifically, zoe refers, in Agamben's work, to the vulnerability of the human body to be reduced to these nonhuman states by the intervention of sovereign power. Zoe is consequently assimilated to death in the sense of the corpse, the liminal bodily existence of a life that does not qualify as human.

Agamben is heir to Heidegger's thought on finitude, to what Agamben calls "bare life" or "the rest" after the humanized "bio-logical" wrapping is taken over (Agamben 1998). "Bare life" is that in you which sovereign power can kill. It is the body as disposable matter in the hands of the despotic force of power (*potestas*). "Bare life" inscribes fluid vitality in the heart of the state system's mechanisms of capture. Agamben is sensitive to the fact that this vitality, or "aliveness," is all the more mortal for that. Referring to the Heideggerian tradition, he stresses the tragic aspects of modernity—the cruelty, violence, wars, destruction, and disruption of traditional ways. Agamben's "bare life" marks the negative limit of modernity and the abyss of totalitarianism that constructs conditions of human passivity.

The position of zoe in Agamben's system is analogous to the role and location of language in psychoanalytic theory: it is the site of constitution, or capture, of the subject. This "capture" functions by positing, as an a posteriori construction, a prelinguistic dimension of subjectivity that is apprehended as "always already" lost and out of reach. Zoe—like Lacan's prediscursive, Kristeva's chora, and Irigaray's maternal feminine—becomes for Agamben the ever-receding horizon of an alterity, which has to be included as necessarily excluded in order to sustain the framing of the subject in the first place. Thus finitude is introduced as a constitutive element within the framework of subjectivity. It also fuels an affective economy of loss and melancholia at the heart of the subject (Braidotti 2002). This view is linked to Heidegger's theory of Being as deriving its force from the annihilation of animal life.

Agamben perpetuates the philosophical habit of taking mortality or finitude as the transhistorical horizon for discussions of "life." The fixation on Thanatos—a fixation that Nietzsche criticized more than a century ago—is still very present in critical debates today. It often produces a gloomy and pessimistic vision, not only of power but also of the technological developments that propel the regimes of biopower. I beg to differ from the habit that favors the deployment of the problem of bios/zoe on the horizon of death, or in the liminal state of not-life, or in the spectral economy of the never-dead. Instead, I prefer to stress the generative powers of zoe by turning to the Spinozist ontology defended by Deleuze and Guattari (1972, 1980).

No reason other than the sterility of habit justifies the emphasis on death as the horizon for discussions about the limits of our understanding of the human. Why not look at the affirmative aspects of exactly the same issue? Speaking from the position of an embodied and embedded female subject, I find the metaphysics of finitude a myopic way of putting the question of the limits of what we call "life." It is not because Thanatos always wins in the end that it should enjoy such conceptual high status. Death is overrated. The ultimate subtraction is, after all, only another phase in a generative process. It is too bad, of course, that the relentless generative powers of death require the suppression of that which is nearest and dearest to me—namely, myself. As psychoanalysis teaches us, it is unthinkable for the narcissistic human subject that life should go on without its own vital being-there. Freud was the first to analyze the blow that death inflicts on the fundamental narcissism of the human subject. The process of confronting the thinkability of a life that may not have "me" or any "human" at its center is actually a sobering and instructive process. This is the very beginning for an ethics of sustainability that aims to shift the focus toward the positivity of zoe.

By contrast with the positioning of zoe as the liminal condition of the living subject—its "becoming corpse," so to speak—I want to borrow freely from Deleuze's (1986b) take on Spinoza in order to think both the positivity of zoe and its being "always already" there. I do so, however, without reference to a linguistic model of interpretation. That model is one that rests on the fundamental rules of metaphor and metonymy. As such, it partakes of and in turn is constituted by the very dialectics of sameness and difference that I am committed to overcoming. Moreover, a linguistic model of interpretation imposes the primacy of a representational way of thinking, which I consider inadequate, given the schizoid and intrinsically nonlinear structure of the global economies of advanced capitalism. As models to account for the kind of subjects we have already become, representational thinking and the linguistic turn are outdated. I opt here instead for a neomaterialist, embodied, embedded approach.

The key to this conceptual shift is the overturning of anthropocentrism as the bottom line of the critique of subjectivity. Poststructuralism initiated that critique by declaring, with Foucault, the "death" of the humanistic subject of knowledge. Nowadays we are experiencing a further stage in this process, and, as the rhizomic philosophies of Deleuze and Guattari point out, we are forced to confront the built-in anthropocentrism that prevents us from relinquishing the categorical divide between bios and zoe and thus makes us cling to the superiority of consciousness in spite of our poststructuralist skepticism toward this very notion. The monist political ontology of Spinoza can rescue us from this contradiction by pushing it to the point of implosion. Through the theory of nomadic becomings or planes of immanence, Deleuze and Guattari (1980) dissolve and reground the subject in an ecophilosophy of multiple belongings. This takes the form of a strong emphasis on the prehuman or even nonhuman elements that compose the web of forces, intensities, and encounters that contribute to the making of nomadic subjectivity. The subject for Deleuze and Guattari is an eco-logical entity.

The term "zoe" refers to the endless vitality of life as a process of continuous becoming. Guattari refers to this process as a transversal form of subjectivity, or "transindividuality." This diffuse yet grounded subject position achieves a double aim: it critiques individualism, and it supports a notion of subjectivity in the sense of qualitative, transversal, group-oriented agency. Lest this mode of subjectivity be mistaken for epistemological anarchy, let me emphasize a number of features of a cartography that takes life as the subject of political discourse.

First, the techno-logical body is in fact an eco-logical unit. This zoe-techno-body is marked by interdependence with its environment, through a structure of mutual flows and data transfers that is best configured by the notion of viral contamination (Ansell-Pearson 1997), or intensive interconnectedness. This nomadic ecophilosophy of belonging is complex and multilayered.

Second, this environmentally bound subject is a collective entity, moving beyond the parameters of classical humanism and anthropocentrism. The human organism is an in-between that is plugged into and connected to a variety of possible sources and forces. As such, it is usefully defined as a machine, which does not mean that it is an appliance or anything with a specifically utilitarian aim but rather that it is something simultaneously more abstract and more materially embedded. The minimalist definition of a body-machine is "an embodied affective and intelligent entity that captures, processes, and transforms energies and forces." Being environmentally bound and territorially based, an embodied entity constantly feeds on, incorporates, and transforms its (natural, social, human, or technological) environment. Being embodied in this high-tech ecological manner means being immersed in fields of constant flows and transformations. Not all of them are positive, of course, although in such a dynamic system this cannot be known or judged a priori.

Third, such a subject of bios/zoe power raises questions of ethical urgency. Given the acceleration in processes of change, how can we tell the difference among the different flows of change and transformations? To answer these questions, I am developing in this chapter a sustainable brand of nomadic ethics. The starting point is the relentless generative force of bios and zoe and the specific brand of transspecies egalitarianism that they establish with the human. The ecological dimension of philosophical nomadism consequently becomes manifest and, with it, its potential ethical impact. It is a matter of forces as well as of ethology.

Fourth, the specific temporality of the subject needs to be rethought. The subject is an evolutionary engine endowed with her or his own embodied temporality, both in the sense of the specific timing of the genetic code and in the sense of the more genealogical time of individualized memories. If the embodied subject of biopower is a complex molecular organism, a biochemical factory of steady and jumping genes, an evolutionary entity endowed with its own navigational tools and a built-in temporality, then we need a form of ethical values and political agency that can reflect this high degree of complexity.

Fifth, and last, this ethical approach cannot be dissociated from considerations of power. The bios/zoe-centered vision of the technologically mediated subject of postmodernity or advanced capitalism is fraught with internal contradictions. Accounting for them is the cartographic task of critical theory, and an integral part of this project is to account for the implications they entail for the historically situated vision of the subject (Braidotti 2002). The bios/zoe-centered egalitarianism potentially conveyed by current technological transformations has dire consequences for the humanistic vision of the subject. The egalitarianism at stake here displaces both the old-fashioned humanistic assumption that "man" is the measure of all things and the anthropocentric idea that the only bodies that matter are human. The vital politics of life as zoe, defined as a generative force, resets the terms of the debate and introduce an ecophilosophy of belonging that includes both species equality and posthumanist ethics.

In other words, the potency of bios/zoe displaces the phallogocentric vision of consciousness, which hinges on the sovereignty of the "I." It can no longer be safely assumed that consciousness coincides with subjectivity, or that either of them is in charge of the course of historical events. Liberal individualism and classical humanism alike are disrupted at their very foundations by the social and symbolic transformations induced by our historical condition. This situation, far from being a mere crisis of values, confronts us with a formidable set of new opportunities. Renewed conceptual creativity and a leap of the social imaginary are needed in order for us to meet the challenge. Classical humanism, with its rationalistic and anthropocentric assumptions, is a hindrance rather than a help in this process. Therefore, as one possible response to this challenge, I propose a posthumanistic brand of nonanthropocentric vitalism.

SUSTAINABLE NOMADIC ETHICS

To defend this position, I start from the concept of a sustainable self that aims at endurance. Endurance has a temporal dimension; it has to do with lasting in time. It is therefore connected to duration and self-perpetuation (traces of Bergson here). But it is also connected to the space of the body as an enfleshed field of actualization of passions or forces. Endurance evolves affectivity and joy (traces of Spinoza), as in the capacity for being affected by these forces to the point of pain or extreme pleasure—each of which comes to the same. It means putting up with hardship and physical pain.

Endurance, apart from providing the key to an etiology of forces, is an ethical principle of affirmation of the positivity of the intensive subject. Endurance is joyful affirmation as *potentia*. The subject is a spatial-temporal compound that frames the boundaries of processes of becoming. This compound works by transforming negative into positive passions through the power of an understanding that is no longer indexed to a phallogocentric set of standards but is to a certain extent unhinged, and therefore affective. Turning the tide of negativity is a transformative process that achieves significant reformulation of the link between understanding and freedom. By introducing a noncognitive idea of understanding, the notion of endurance suggests freedom of understanding through the awareness of our limits, and hence also of our relative bondage. This transformation results in the freedom to affirm one's essence as joy, by encountering and mingling with other bodies, entities, beings, and forces. Ethics means faithfulness to this *potentia*, which is my definition of the desire to become. Desire, here, is ontological and not erotic;

it is the desire to be (or rather to become) and not the desire to have. Here, the verb "to become" indicates an open-ended process and not one whose goal is a specific entity, one bounded into its own being. The process of desire is driven by affect. Affectivity can intrinsically be understood as positive.

This does not mean, however, that it is uncontaminated by the impact of the specific political economy of desire implemented by advanced capitalism—quite the contrary. It is embedded in it, so as to provide a forceful antidote to it. For instance, contemporary culture tends to react to technological advances with a double pull that swings from hype to nostalgia, or from euphoria to melancholia. This affective economy sets the mood for the psychopathologies of today. A nomadic ethics of affirmation pleads instead for a sober form of lucidity that aims at sustainable transformations. It avoids references to the paradigms of human nature (be it a biological or psychic paradigm or a paradigm of genetic essentialism), and hence to the fear of moral relativism, while accounting for the fact that bodies have indeed become biotechnocultural constructs immersed in networks of complex, simultaneous, and potentially conflicting power relations.

Another example of the complex relationship of affectivity to advanced capitalism concerns the becoming-woman of labor. A system that prides itself on being an information society is actually based on immaterial labor, which involves communication, cooperation, data processing, information management, and media work. This labor force trades phonetic skills, health and good looks, linguistic ability, and proper language, and it accents services as well as attention and great concentration. Consequently, it prioritizes the production and reproduction of affects, such as caring, serviceability, and the re-creation of fast-disappearing community bonds. Historically, this has been women's work, constituting a central piece of capitalist production. This analysis is offered, notably, by Hardt and Negri's (2000) critique of globalization, but they do not think this problem through to the structures of the gender politics of advanced capitalism or to the specific contradictions inherent in the process of the feminization of labor. Contrary to the metaphysics of labor proposed by Hardt and Negri, nomadic politics argues for a more grounded approach. The digital workers of the new global economy express an acute and explicit awareness of their location in space and time. Therefore, they raise serious questions, not only about the affective elements of their labor but also about its material grounds, which entail border crossings, shifts in mobility, and paths of deterritorialization. It is quite clear that the allegedly ethereal nature of cyberspace, and the flow of global capital's mobility that it sustains, are fashioned by the material labor of real living bodies from and/or in areas of the world that are thought to be peripheral. Thus this space of capital fluctuation is racialized and sexualized to a very high degree. A new "feminization" of the virtual workforce has taken place, and with it has come a deterioration in rights and in conditions. What is needed to account for it is not the euphoric and at times hyperbolic language of neo-Marxism but rather the embedded and embodied brands of materialism that feminist theory has developed. There is no need for an overarching metanarrative of one revolutionary multitude if one is working with feminist notions of situated knowledge (Haraway 1988) or my own nomadic philosophy of radical immanence (Braidotti 2002, 2006).

Affectivity is the force that aims to fulfill the subject's capacity for interaction and freedom. Affectivity is Spinoza's central notion of desire as *conatus*, or the drive to fulfill one's essential inner freedom through processes of becoming. This is linked to the notion of *potentia* as the affirmative aspect of power. In a neo-Spinozist perspective, *conatus* is implicitly positive in that it expresses the essential best of the subject. The subject is joyful and pleasure-prone, and it is immanent in that it coincides with the terms and modes of its expression. This means, concretely, that ethical behavior confirms, facilitates, and enhances the subject's *potentia*—the capacity to express her or his freedom. The positivity of this desire to express one's innermost and constitutive freedom is conducive to ethical behavior. Nevertheless, it leads to ethical behavior only if the subject is capable of making the positivity of desire last and endure, thus allowing it to sustain its own *potentia*. Unethical behavior achieves quite the opposite result: it denies, hinders, and diminishes that *potentia*. Thus unethical behavior is unable to sustain becoming.

This introduces a temporal dimension into the discussion and leads to the very conditions of possibility for the future—to futurity as such. For an ethics of sustainability, the expression of positive affects is what causes the subject to last or endure. Expression of positive affects is like a long-lasting source of energy at the affective core of subjectivity. To better understand the importance of temporality for a sustainable nomadic ethics, we must turn again to Deleuze's nomadology. In my view, his nomadology is a philosophy of immanence, which rests on the idea of sustainability as a principle of containment and tolerable development of a subject's resources. Those resources can be environmental, affective, or cognitive. A subject thus constituted inhabits a time that is the active tense of continuous becoming. Deleuze (1966, 1988) defines the latter with reference to Bergson's concept of duration, thus proposing the notion of the subject as an entity that lasts, that is to say, an entity that endures sustainable changes and transformation and enacts them around itself in a community or collectivity. Deleuze (1968b) disengages the notion of endurance from the metaphysical tradition that associates it with the idea of essence, and hence also of permanence. He injects endurance with spatial-temporal force. It can be seen as a form of transcendental empiricism, or of antiessentialist vitalism. From this perspective, even the Earth (Gaia) is posited as a partner in a community that is still to come, one to be constructed by subjects who will interact with the Earth differently. This idea is in some ways close to "deep ecology" (Naess 1977) but is radically antiessentialist in its understanding of the structure and location of the human within it.

What, then, is this sustainable subject? It is a slice of living, sensible matter—a self-sustaining system activated by a fundamental drive toward life, *potentia* rather than *potestas*. It is neither something activated by the will of God nor the secret encryption of the genetic code, yet this subject is psychologically embedded in the corporeal materiality of the self. The enfleshed intensive or nomadic subject is an in-between: a folding in of external influences, and a simultaneous folding out of affects. As a mobile entity—mobile in space and time—this subject is continually in process but is also capable of lasting through sets of discontinuous variations while remaining extraordinarily faithful to itself.

Faithfulness to oneself is not to be understood in the mode of psychological or sentimental attachment to a personal identity that often is little more than a social security number and a set of photo albums (or, as José van Dijck shows in chapter 8, a digital weblog). Nor is it the mark of authenticity of a self—"me, myself, and I"—that is a clearinghouse for narcissism and paranoia, the great pillars on which Western identity predicates itself. Rather, it is the faithfulness of mutual sets of interdependence and interconnections. The sustainable subject is made up of sets of relations and encounters. These multiple relationships encompass all levels of one's multilayered subjectivity, binding the cognitive to the emotional, the intellectual to the affective, and connecting them all to a socially embedded ethics of sustainability. Thus the faithfulness at stake in nomadic ethics coincides with the awareness of one's condition of interaction with others—in other words, with one's capacity to affect and to be affected. Transposed to a temporal scale, this is the faithfulness of duration, the expression of one's continuing attachment to certain dynamic spatial-temporal coordinates. To be faithful to oneself is to endure.

In a philosophy of temporally inscribed radical immanence, subjects differ, but they differ along materially embedded coordinates; they come in different mileages, temperatures, and beats. One can and does shift gears across these coordinates, but one cannot claim all of them all of the time. The latitudinal and longitudinal forces that structure the subject have limits of sustainability. By "latitudinal forces," Deleuze and Guattari (1980) mean the affects of which a subject is capable, according to degrees of intensity or potency—how intensely these affects run. The term "longitudinal forces" means the span of their extension—how far these affects can go. Sustainability has to do with how much a subject can take. Ethics can be understood as a geometry of how much bodies are capable of.

What, then, is this threshold, and how does it get fixed? A radically immanent intensive body is an assemblage of forces, or of flows, intensities, and passions, that solidify (in space) and consolidate (in time) within the singular configuration commonly known as an "individual" self. This intensive and dynamic entity does not coincide with the enumeration of inner rationalist laws, nor is it merely the unfolding of genetic data and information encrypted in the material structure of the embodied self. It is, rather, a portion of the forces just described that is stable enough to sustain and undergo constant, though nondestructive, fluxes of transformation.

It is the body's degrees and levels of affectivity that determine the modes of differentiation. How much vitality or positive power of life is a body capable of? Joyful or positive passions and the transcendence of reactive affects are the desirable mode of affirmation of the specific portion of "life" that one happens to be. The emphasis on lived existence implies a commitment to duration and, conversely, a rejection of self-destruction. Positivity is built into this program through the idea of thresholds of sustainability. Thus an ethically empowering option increases one's *potentia* and creates joyful energy in the process. The conditions that can encourage such a quest are not just historical; but also concern processes of self-transformation or self-fashioning. Because all subjects share in this common nature, there is a common ground on which to negotiate mutual interests as well as eventual conflicts.

Only through empirical experimentation can one know whether one has reached the threshold of sustainability. Sustainable ethics is a process, not a moral imperative. This is where the nonindividualistic vision of the subject as embodied, and hence affective and interrelational, is of major consequence. Your body will tell you if and when you have reached a threshold or a limit. The warning can take the form of your body's opposing resistance through illness, feelings of nausea, or somatic manifestations like fear, anxiety, or a sense of insecurity. Whereas the semiotic-linguistic frame of psychoanalysis reduces these manifestations to symptoms awaiting interpretation, I see them as corporeal warning signals or boundary markers that express a clear message: *Too much!* One reason why Deleuze and Guattari are so interested in studying self-destructive or pathological modes of behavior, such as schizophrenia, masochism, anorexia, various forms of addiction, and the black hole of murderous violence, is precisely because they want to explore their function as markers of thresholds. This project assumes a qualitative distinction between, on the one hand, the desire that propels the subject's expression of its *potentia* and, on the other hand, the constraints imposed by society. The specific, contextually determined conditions are the forms in which the desire is actualized or actually expressed. The thresholds of sustainability need to be spelled out through experiments, which are necessarily relational and occur in encounters with others. To understand these interactive and affective "bits of life," we need new cognitive and sensory mappings of the thresholds of sustainability for bodies that are in the process of transformation.

In order to sustain interconnections and interrelations, the subject needs to develop some form of self-knowledge, which cannot be reduced to mere cognition. Affectivity is an essential part of this intensive notion of self-knowledge, which itself is driven by desire and by *potentia.* Understanding is like mapping thresholds of becoming, and hence of sustainability. It involves self-preservation, not in the liberal individualistic sense of the term but as the actualization of one's essence, that is to say, of one's ontological drive to become. This is neither an automatic nor an intrinsically harmonious process, insofar as it involves interconnection with other forces and consequently also conflicts and clashes. Negotiations have to occur, and to serve as stepping-stones to sustainable flows of becoming. The bodily self's interaction with its environment can either increase or decrease that body's *conatus* or *potentia.* The mind, as a sensor that prompts understanding, can assist by helping the bodily self to discern and choose those forces that increase its power of acting and its activity in both physical and mental terms. A higher form of self-knowledge, through an understanding of the nature of one's affectivity, is the key to a Spinozist ethics of empowerment. It includes a more adequate understanding of the interconnections between the self and a multitude of other forces, thus undermining the liberal individual understanding of the subject. It also implies, however, the body's ability to comprehend and physically sustain a greater number of complex interconnections and to deal with complexity without becoming overburdened. Therefore, only an appreciation of increasing degrees of complexity can guarantee the freedom of the mind in the awareness of its true, affective, dynamic nature.

At this point, it is important to stress that sustainability is about decentering anthropocentrism. The ultimate implication is a displacement of the human in the new, complex compound of highly generative posthumanities. In my view, the sustainable subject has a nomadic subjectivity because the notion of sustainability brings together ethical, epistemological, and political concerns under cover of a nonunitary vision of the subject. Let's not pretend, however, that displacement of anthropocentrism is easy. "Life" privileges assemblages of a heterogeneous kind. Animals, insects, machines are as many fields of forces or territories of becoming. The life in me is not only, not even, human.

Far from precipitating us into an abyss of amorality and nihilism, this approach fosters the possibility that more situated forms of interaction and microuniversals will emerge. Contemporary science and biotechnologies affect the very fiber and structure of the living, creating a negative unity among humans. The Human Genome Project, for instance, unifies the entire human species in the urgency to organize an opposition to commercially owned, profit-minded genetic technologies. Franklin, Lury, and Stacey (2000: 26) refer to "panhumanity," by which they mean a global sense of interconnection between the human and the nonhuman environment as well as among the different subspecies within each category, interconnections that create a web of intricate interdependences. Most of this mutual dependence is of the negative kind, viewed in terms of "a global population at shared risk of global environmental destruction and united by collective global images" (ibid.). Nevertheless, this form of postmodern human interconnection also has positive elements. Franklin and her colleagues argue that this universalization is one of the effects of the global economy and is part of a recontextualization of the market economy that is currently under way. They also describe it in Deleuzian terms, as "unlimited finitude" or "visualization without horizon," and see it as a potentially positive source of resistance.

The paradox of this new panhumanity involves not only a sense of shared and associated risks but also pride in technological achievements and in the wealth that comes with them. On a more positive note, there is no doubt that we are all in this together. Any nomadic philosophy of sustainability worthy of its name will have to start from this assumption and reiterate it as a fundamental value. The point, however, is to define the part called "we" part and the content of "this"—that is to say, the community in relation to singular subjects, and the norms and values of a political ecophilosophy of sustainability. The debates on these issues, in fields as diverse as environmental, political, social, and ethical theory (Becker and Johan 1999), show a range of potentially contradictory positions. From the "world governance idea" (Brundtland Commission 1987) to the ideal of a "world ethos" (Kung 1998) through a large variety of ecological brands of feminism, the field is wide open. In other words, we are witnessing a proliferation of locally situated universalist claims. Far from being a symptom of relativism, the proliferation of these claims asserts the radical immanence of the subject. They constitute the starting point for a web of intersecting forms of situated accountability—that is to say, an ethics. The whole point is to elaborate sets of criteria for a new ethical system that is still to be brought into being , and that will steer a course between humanistic nostalgia and neoliberal euphoria. In my view, this can only be an ethics that takes life (as bios and as zoe) as its point of reference, not for the sake of restoring unitary norms or celebrating the master narrative of global profit, but for the sake of sustainability.

CONCLUSION

I hope to have shown that we inhabit the paradoxes of biopower in technologically mediated societies and therefore need new ethics, cosmologies, and worldviews that are appropriate to our high level of technological development and to the global issues that are connected with it. For this purpose, I began in this chapter to develop a sustainable nomadic ethics (see also Braidotti 2006). Far from pointing to the residual mysticism of a notion of life as vital holism, it is meant to be a concrete plan for embedding new figurations of living subjectivities in the posthumanist mode. This is an evolutionary tale of the nondeterministic type, bypassing quantitative multiplication to achieve a qualitative leap of values. These values do not correspond a priori to established moral conventions; rather, they evolve alongside political analyses that do justice to the sets of ferocious structural injustices and insidious modes of dispossession that mark the global economy. Therefore, they include serious analyses of power relations.

Bios/zoe power is a political economy that distributes entitlements to death as well as to survival. Consequently, we need cultural, spiritual, and ethical values, whether myths, narratives, or representations, that are adequate to this new civilization we inhabit. The merger of the human with the technological in a machinic environment, not unlike the symbiotic relationship between the animal and its habitat, results in a new compound, a new kind of open whole. This is neither a holistic fusion nor a Christian form of transcendence. Rather, I have stressed the materialist plane of radical immanence. This in-between-ness is best addressed, not as biology, and certainly not as bioethics, but as an ethology of forces, by which I mean an ethics of mutual interdependence and of sustainable interactions. More creativity is needed to refigure these ethical interconnections. Instead of falling back on sedimented habits of thought, I have proposed a leap forward into the complexities and paradoxes of our time. Whatever figuration of a new biocentered humanity we may be able to agree on, it can only be a temporary and hybrid mixture. Bios/zoe power keeps the "human" hung up between a future that cannot provide a safe guarantee and a fast rate of current change that demands one. This tantalizing loose end expresses the perverse logic of biopower as a regime that points to possible futures while blocking and controlling access to them in such a way as to ensure that "life" never reaches the higher levels of intensity of which it is potentially capable.

Positive metamorphosis can be seen as political passion. It endorses the kinds of becomings that destabilize dominant power relations and deterritorialize fixed identities and mainstream values. Such a metamorphosis infuses a joyful sense of empowerment into a subject that is always in the process of becoming. This passion is ethical as well as political because it mobilizes the critical resources of the intellect as well as the creative imagination for the cause of human freedom as a collectively held hope.

NOTES

1. For a more detailed and critical overview of inflationary discourses around the concept of life, see Braidotti (2006).

2. Elsewhere, I have developed the notion of the nomadic subject as a materially embodied, historically embedded cartography of subjectivity embodying a set of multiple, complex, and internally contradictory relations (Braidotti 1994, 2002, 2006). As a feminist notion, nomadic subjectivity relates both to sexual difference, as a political project of empowering a virtual feminine, and to feminist activism. Nomadic subjectivity is a philosophy of immanence or active becoming. It relies on a Spinozist political ontology, which provides grounding for ethical as well as political accountability, against postmodern fragmentation, on the one hand, and tragic masculine celebrations of "bare life" as ontological lack, on the other. My nomadic subject is in dialogue with other figurations of mobility and displacement in contemporary critical theory and in postcolonial and migration theories, specifically addressing the predicament of a critique of Eurocentrism from within.

3. It is interesting to note that in the proliferating discourses on life (and consequently on death as well), eros does not receive much attention. The scope of this chapter does not allow me to explore whether these are indeed "eros-less" times.

Bibliography

Aarseth, Espen J. 1997. *Cybertext: Perspectives on Ergodic Literature.* Baltimore, Md.: Johns Hopkins University Press.

Agamben, Giorgio. 1998. *Homo Sacer: Sovereign Power and Bare Life.* Stanford, Calif.: Stanford University Press.

Ansell-Pearson, Keith. 1997. *Viroid Life: Perspectives on Nietzsche and the Transhuman Condition.* London: Routledge.

Armitt, Lucie, ed. 1990. *Where No Man Has Gone Before.* London: Routledge.

Aronowitz, Stanley, and Henry A. Giroux. 1993. *Education Still Under Siege.* Westport, Conn.: Bergin and Garvey.

Aronowitz, Stanley, et al., eds. 1996. *Technoscience and Cyberculture.* New York: Routledge.

Baert, Renee. "Lifeworlds." 2001. In Wendy Kirkup, *Echo* (exhibition catalog with CD-ROM). Newcastle: Locus+.

Baird, Nicola. "Body of Evidence." 2002. *The Guardian*, Sept. 25.

Bal, Mieke, Jonathan Crew, and Leo Spitzer, eds. 1999. *Acts of Memory: Cultural Recall in the Present.* Hanover, N.H.: University Press of New England.

Balsamo, Ann. 1996. *Technologies of the Gendered Body: Reading Cyborg Women.* Durham, N.C.: Duke University Press.

Barad, Karen. 1998. "Getting Real: Technoscientific Practices and the Materialization of Reality." *Differences: A Journal of Feminist Cultural Studies* 10(2): 87–126.

———. 2003. "Posthumanist Performativity: How Matter Comes to Matter." *Signs: Journal of Women in Culture and Society* 28(3): 801–31.

———. 2007. *Meeting the Universe Halfway: Quantum Physics and the Entanglement of Matter and Meaning.* Durham, N.C.: Duke University Press.

Barker, Martin. 1981. *The New Racism: Conservatives and the Ideology of the Tribe.* London: Junction Books.

Barr, Marlene, ed. 1981. *Future Females: A Critical Anthology.* Bowling Green, Ohio: Bowling Green State University Popular Press.

Becker, Egon, and Thomas Johan, eds. 1999. *Sustainability and the Social Sciences: A Cross-Disciplinary Approach to Integrating Environmental Considerations into Theoretical Reorientation.* London: Zed Books/UNESCO.

Beer, Gillian. 1983. *Darwin's Plots: Evolutionary Narrative in Darwin, George Eliot, and Nineteenth-Century Fiction.* London: Routledge & Kegan Paul.

———. 2000. *Darwin's Plots: Evolutionary Narrative in Darwin, George Eliot, and Nineteenth-Century Fiction*, 2nd ed. New York: Cambridge University Press.

Bennett, M. R., and P. M. S. Hacker. 2003. *Philosophical Foundations of Neuroscience.* Oxford: Blackwell.

Birke, Lynda. 1999. *Feminism and the Biological Body.* New Brunswick, N.J.: Rutgers University Press.

Bohr, Niels. 1958. *Atomic Physics and Human Knowledge.* New York: Wiley.

Bolter, Jay David. 1991. *Writing Space: The Computer, Hypertext, and the History of Writing*. Hillsdale, N.J.: Lawrence Erlbaum.

———, and Richard Grusin. 2000. *Remediation: Understanding New Media*. Cambridge, Mass.: MIT Press.

Booth, Austin. 2002. "Women's Cyberfiction: An Introduction." In Mary Flanagan and Austin Booth, eds., *Reload: Rethinking Women + Cyberculture*. Cambridge, Mass.: MIT Press.

Borell, Merriley. 1985. "Organotherapy and the Emergence of Reproductive Endocrinology." *Journal of the History of Biology* 18(1): 1–30.

Boston Women's Health Book Collective. 1971. *Our Bodies, Ourselves*. New York: Simon and Schuster.

Braidotti, Rosi. 1994. *Nomadic Subjects*. New York: Columbia University Press.

———. 2002. *Metamorphoses: Towards a Materialist Theory of Becoming*. Cambridge: Polity Press.

———. 2006. *Transpositions: On Nomadic Ethics*. Cambridge: Polity Press.

———, J. Nieboer, and S. Hirs, eds. 2002. *The Making of European Women's Studies*, vol. 4. Utrecht: ATHENA (Advanced Thematic Network in Activities in Women's Studies in Europe), Utrecht University.

Brugh, Marcel aan de. 2000. "Moderne Voortplantingstechnieken Verstoren Embryonale Ontwikkeling." *NRC Handelsblad*, Jan. 22.

Brundtland Commission. 1987. *Our Common Future*. Oxford: Oxford University Press.

Bruno, Giuliana. 1992. "Spectorial Embodiments: Anatomies of the Visible and the Female Bodyscape." In Paula Treichler and Lisa Cartwright, eds., *Imaging Technologies, Inscribing Science*. Special issue of *Camera Obscura* 28: 239–61.

———. 1993. *Streetwalking on a Ruined Map: Cultural Theory and the City Films of Elvira Notari*. Princeton, N.J.: Princeton University Press.

Bryld, Mette, and Nina Lykke. 2000. *Cosmodolphins: Feminist Cultural Studies of Technology, Animals and the Sacred*. London: Zed Books.

Bukatman, Scott. 1993. *Terminal Identity: The Virtual Subject in Postmodern Science Fiction*. Durham, N.C.: Duke University Press.

Burley, Helen. 2002. "Stevie Wonder." *Earth-Matters: Friends of the Earth Supporter Magazine* 53: 25.

Bush, Vannevar. 1945. "As We May Think." *Atlantic Monthly*, July: 101–8. Online at http://www.theatlantic.com/unbound/flashbks/computer/bushf.htm (retrieved Feb. 21, 2007).

Butler, Judith. 1993. *Bodies That Matter: On the Discursive Limits of "Sex."* New York: Routledge.

Cadbury, Deborah. 1998. *The Feminization of Nature: Our Future at Risk*. London: Penguin Books.

Canguilhem, Georges. 1994. *The Vital Rationalist: Selected Writings from Georges Canguilhem*. New York: Zone Books.

Cartwright, Lisa. 1992. "Women, X-rays, and the Public Culture of Prophylactic Imaging." In Paula Treichler and Lisa Cartwright, eds., *Imaging Technologies, Inscribing Science*. Special issue of *Camera Obscura* 29: 19–54.

———. 1995. *Screening the Body: Tracing Medicine's Visual Culture*. Minneapolis: University of Minnesota Press.

Cattell, Emma. 2003. "What's for Dinner, Darling? Ooh, Endocrine Disrupters—My Favourite." *Independent on Sunday*, March 30.

Cavallaro, Dani. 2000. *Cyberpunk and Cyber-*

culture: Science Fiction and the Work of William Gibson*. London: Athlone Press.

Centraal Bureau voor de Statistiek. 1999a. *Vademecum Gezondheidsstatistiek Nederland 1999*. Voorburg: CBS.

———. 1999b. *Vital Events: Past, Present, and Future of the Dutch Population*. Voorburg: CBS.

Cheetham, Mark, and Elizabeth D. Harvey. 2002. "Obscure Imaginings: Visual Culture and the Anatomy of Caves." *Journal of Visual Culture* 1(1): 105–26.

Churchland, Paul. 1984. *Matter and Consciousness: A Contemporary Introduction to the Philosophy of Mind*. Cambridge, Mass.: MIT Press.

Churchland Smith, Patricia. 1986. *Neurophilosophy: Toward a Unified Science of the Mind-Brain*. Cambridge, Mass.: MIT Press.

Clarke, Arthur C. 1964. *Profiles of the Future*. London: Pan Books.

Clifford, James, and George Marcus, eds. 1986. *Writing Culture: The Poetics and Politics of Ethnography*. Berkeley: University of California Press.

Clynes, Manfred E., and Nathan S. Kline. 1995. "Cyborgs and Space." In Chris Hables Gray, ed., *The Cyborg Handbook*. London: Routledge.

Colborn, Theo, Dianne Dumanoski, and John Peterson. 1996. *Our Stolen Future: Are We Threatening Our Fertility, Intelligence, and Survival?* New York: Dutton.

Collins, Patricia Hill. 1998. "It's All in the Family: Intersections of Gender, Race, and Nation." *Hypatia* 3: 62–82.

Creed, Barbara. 1993. *The Monstrous Feminine: Film, Feminism and Psychoanalysis*. London: Routledge.

Cussins, Charis. 1998. "Producing Reproduction: Techniques of Normalization and Naturalization in Infertility Clinics." In Sarah Franklin and Hellen Ragoné, eds., *Reproducing Reproduction: Kinship, Power, and Technological Innovation*. Philadelphia: University of Pennsylvania Press.

Danet, Brenda. 1997. "Books, Letters, Documents: The Changing Aesthetics of Texts in Late Print Culture." *Journal of Material Culture* 2: 5–38.

Davis-Floyd, Robbie, and Joseph Dumit, eds. 1998. *Cyborg Babies: From Techno-Sex to Techno-Tots*. New York: Routledge.

Dawkins, Richard. 1976. *The Selfish Gene*. Oxford: Oxford University Press.

Dawson, Graham. 1994. *Soldier Heroes: British Adventure, Empire and the Imagining of Masculinities*. London: Routledge.

De Bendern, Paul. 2002. "Toxins Put Arctic Polar Bears and Humans at Risk." Online at http://www.planetark.org/dailynews story.cfm/newsid/18010/story.htm (retrieved Feb. 21, 2007).

De Lauretis, Teresa. 1990. "Upping the Anti (sic) in Feminist Theory." In Marianne Hirsch and Evelyn Fox Keller, eds., *Conflicts in Feminism*. New York: Routledge.

Deleuze, Gilles. 1962. *Nietzsche et la philosophie*. Paris: Presses Universitaires de France.

———. 1966. *Le bergsonisme*. Paris: Presses Universitaires de France.

———. 1968a. *Différence et répétition*. Paris: Presses Universitaires de France.

———. 1968b. *Spinoza et le problème de l'expression*. Paris: Minuit.

———. 1969. *Logique du sens*. Paris: Minuit.

———. 1983. *Nietzsche and Philosophy*. Trans. Hugh Tomlinson. New York: Columbia University Press.

———. 1988. *Bergsonism*. Trans. Hugh Tomlinson and Barbara Habberjam. New York: Zone Books.

———. 1990a. *Expressionism in Philosophy:*

Spinoza. Trans. Martin Joughin. New York: Zone Books.

———. 1990b. *The Logic of Sense.* Ed. Constantin V. Boundas. Trans. Mark Lester with Charles Stivale. New York: Columbia University Press.

———. 1994. *Difference and Repetition.* Trans. Paul Patton. London: Athlone.

———, and Félix Guattari. 1972. *L'anti-Oedipe: Capitalisme et schizophrénie,* vol. 1. Paris: Minuit.

———. 1977. *Anti-Oedipus: Capitalism and Schizophrenia.* Trans. Robert Hurley, Mark Seem, and Helen R. Lane. New York: Viking Press.

———. 1980. *Mille plateaux: Capitalisme et schizophrénie,* vol. 2. Paris: Minuit.

———. 1987. *A Thousand Plateaus: Capitalism and Schizophrenia.* Trans. Brian Massumi. Minneapolis: University of Minnesota Press.

———. 1991. *Qu'est-ce que la philosophie?* Paris: Minuit.

———. 1994. *What Is Philosophy?* Trans. Hugh Tomlinson and Graham Burchell. New York: Columbia University Press.

Dery, Mark. 1996. *Escape Velocity: Cyberculture at the End of the Century.* New York: Grove Press.

Desmond, Adrian J. 1984. *The Politics of Evolution: Morphology, Medicine, and Reform in Radical London.* Chicago: University of Chicago Press.

———, and James Moore. 1991. *Darwin.* London: Michael Joseph.

Dijck, José van. 1998. *Imagenation: Popular Images of Genetics.* New York: New York University Press.

———. 2004. "Mediated Memories: Personal Cultural Memory as Object of Cultural Analysis." *Continuum: Journal for Media and Cultural Studies* 18(2): 261–77.

———. 2005a. "From Shoebox to Performative Agent: The Computer as Personal Memory Machine." *New Media and Society* 7(2): 291–312.

———. 2005b. *The Transparent Body: A Cultural Analysis of Medical Imaging.* Seattle: University of Washington Press.

———. 2007. *Mediated Memories in the Digital Age.* Stanford, Calif.: Stanford University Press.

Doane, Mary Ann. 2000. "Technophilia: Technology, Representation, and the Feminine." In Gill Kirkup, Linda Janes, Kathryn Woodward, and Fiona Hovenden, eds., *The Gendered Cyborg: A Reader.* London: Routledge.

Donawerth, Jane. 1997. *Frankenstein's Daughters: Women Writing Science Fiction.* New York: Syracuse University Press.

———, and Carole A. Kolmerten, eds. 1994. *Utopian and Science Fiction by Women: Worlds of Difference.* Liverpool: Liverpool University Press.

Douglas, J. Yellowlees. 2001. *The End of Books—or Books without End?* Ann Arbor: University of Michigan Press.

Douglas, Susan. 1999. *Listening In: Radio and the American Imagination.* New York: Random House.

Draaisma, Douwe. 2002. *Waarom het leven sneller gaat als je ouder wordt.* Groningen: Historische Uitgeverij.

Duden, Barbara. 1993. *Disembodying Women.* Cambridge, Mass.: Harvard University Press.

Dutch Health Council. 2001. *Nuclear Transplantation in Cases of Mutations in Mitochondrial DNA.* The Hague: Dutch Health Council.

"Een Kind van Twee Moeders? Niet Echt." 2001. *Trouw,* May 7.

Emmeche, Claus. 1994. "Is Life a Multiverse Phenomenon?" In Christopher G. Langton, ed., *Artificial Life III* (proceedings of the Workshop on Artificial Life, Santa Fe, N.M., June 1992). Reading, Mass.: Addison-Wesley.

European Environmental Bureau, European Public Health Alliance Environment Network, Friends of the Earth Europe, Greenpeace International, World Wildlife Fund, and Women in Europe for a Common Future. 2005. "NGOs' Five Key Demands to Improve REACH." Online at http://www.foeeurope.org/safer_chemicals/five_key_demands.pdf (retrieved Feb. 21, 2007).

Fausto-Sterling, Anne. 2000. *Sexing the Body: Gender Politics and the Construction of Sexuality*. New York: Basic Books.

Ferreira, Maria Aline. 2000. "Shelley Jackson's *Patchwork Girl* and Angela Carter's *The Passion of New Eve*: A Comparative Reading." In Álvaro Pina, João Ferreira Duarte, and Maria Helena Serôdio, eds., *Do Esplendor na Relva: Élites e Cultura Comum de Expressão Inglesa*. Lisbon: Edições Cosmos. Online at http://www.cyberartsweb.org/cpace/ht/pg/ferreira.html (retrieved Feb. 21, 2007).

Flanagan, Mary, and Austin Booth, eds. 2002. *Reload: Rethinking Women + Cyberculture*. Cambridge, Mass.: MIT Press.

Foucault, Michel. 1972. *The Archaeology of Knowledge and the Discourse on Language*. Trans. A. M. Sheridan Smith. New York: Pantheon.

———. 1976. *Histoire de la sexualité*, vol. 1: *La volonté de savoir*. Paris: Gallimard.

———. 1978. *The History of Sexuality*, vol. 1. Trans. Robert Hurley. New York: Pantheon.

Franklin, Sarah. 1991. "Fetal Fascinations: New Dimensions to the Medical-Scientific Construction of Fetal Personhood." In Sarah Franklin, Celia Lury, and Jackie Stacey, eds., *Off-Centre: Feminism and Cultural Studies*. London: HarperCollins Academic.

———. 1997. *Embodied Progress: A Cultural Account of Assisted Reproduction*. London: Routledge.

———. 2000. "Life Itself." In Sarah Franklin, Celia Lury and Jackie Stacey, *Global Nature, Global Culture*. London: Sage.

———. 2002. "Flat Life: Conception after Dolly." Paper presented at the Biotechnology, Philospohy and Sex conference, Ljubljana, Slovenia, Oct. 12.

———. 2003. "Ethical Biocapital: New Strategies of Stem Cell Culture." In Sarah Franklin and Margaret Lock, eds., *Remaking Life and Death: Toward an Anthropology of the Biosciences*. Sante Fe, N.M.: School of American Research Press.

———, Celia Lury, and Jackie Stacey, eds. 1991. *Off-Centre: Feminism and Cultural Studies*. London: HarperCollins Academic.

———, and Hellen Ragoné, eds. 1998. *Reproducing Reproduction: Kinship, Power, and Technological Innovation*. Philadelphia: University of Pennsylvania Press.

———, Celia Lury, and Jackie Stacey. 2000. *Global Nature, Global Culture*. London: Sage.

Friedberg, Ann. 1993. *Window Shopping: Cinema and the Postmodern*. Berkeley: University of California Press.

Friends of the Earth. 2006. "Chemical Reaction: Reach for a Toxics-Free Future." Online at http://www.foe.co.uk/campaigns/safer_chemicals/chemical_reaction/news_en.html (retrieved Feb. 21, 2007).

——— and Natural Childbirth Trust. 2001. *Chemicals in the Home: A Parents' Guide*. London: Friends of the Earth and National Childbirth Trust.

Gibson, William. 1984. *Neuromancer*. New York: Ace.

Gilbert, Sandra M., and Susan Gubar. 1979. *The Madwoman in the Attic: The Woman Writer and the Nineteenth-Century Literary Imagination*. New Haven, Conn.: Yale University Press.

Ginsburg, Faye D., and Rayna Rapp, eds. 1995. *Conceiving the New World Order: The Global Politics of Reproduction*. Berkeley: University of California Press.

Gray, Ann. 2003. *Research Practice for Cultural Studies: Ethnographic Methods and Lived Culture*. London: Sage.

Griffin, Gabrielle, and Rosi Braidotti, eds. 2002. *Thinking Differently: A Reader in European Women's Studies*. London: Zed Books.

Gross, David. 2000. *Lost Time: On Remembering and Forgetting in Late Modern Culture*. Amherst: University of Massachusetts Press.

Gross, Paul R., and Norman Levitt. 1994. *Higher Superstition: The Academic Left and Its Quarrels with Science*. Baltimore, Md.: Johns Hopkins University Press.

Grossberg, Lawrence, Cary Nelson, and Paula Treichler, eds. 1992. *Cultural Studies*. London: Routledge.

Guattari, Felix. 1995. *Chaosmosis: An Ethicoaesthetic Paradigm*. Sydney: Power Publications.

Guisti, R. M., K. Iwamoto, and E. E. Hatch. 1995. "Diethylstilbestrol Revisited: A Review of the Long-Term Health Effects." *Annals of Internal Medicine* 122(10): 778–88.

Haraway, Donna. 1988. "Situated Knowledges: The Science Question in Feminism as a Site of Discourse on the Privilege of Partial Perspective." *Feminist Studies* 14(3): 575–99.

———. 1989. *Primate Visions: Gender, Race, and Nature in the World of Modern Science*. New York: Routledge.

———. 1991a. "A Cyborg Manifesto: Science, Technology, and Socialist-Feminism in the Late Twentieth Century." In Donna Haraway, *Simians, Cyborgs, and Women: The Reinvention of Nature*. London: Free Association Books.

———. 1991b. *Simians, Cyborgs and Women: The Reinvention of Nature*. London: Free Association Books.

———. 1992. "The Promises of Monsters: A Regenerative Politics for Inappropriate/d Others." In Lawrence Grossberg, Cary Nelson, and Paula Treichler, eds., *Cultural Studies*. London: Routledge.

———. 1994. "A Game of Cat's Cradle: Science Studies, Feminist Studies, Cultural Studies." *Configurations* 2(1): 59–71.

———. 1997. *Modest_Witness@Second_Millennium.FemaleMan©_Meets_OncoMouse™: Feminism and Technoscience*. New York: Routledge.

———. 2000. *How Like a Leaf: An Interview with Thyrza Nichols Goodeve*. London: Routledge.

———. 2003. *The Companion Species Manifesto*. Chicago: Prickly Paradigm Press.

———. 2004. *The Haraway Reader*. London: Routledge.

Harding, Sandra. 1991. *Whose Science? Whose Knowledge? Thinking from Women's Lives*. Ithaca, N.Y.: Cornell University Press.

Hardt, Michael, and Antonio Negri. 2000. *Empire*. Cambridge, Mass.: Harvard University Press.

Harter, Deborah A. 1996. *Bodies in Pieces: Narratives and the Poetics of the Fragment*. Stanford, Calif.: Stanford University Press.

Hartouni, Valerie. 1992. "Fetal Exposures: Abortion Politics and the Optics of Allusion." In Paula Treichler and Lisa Cartwright, eds., *Imaging Technologies, Inscribing Science*. Special issue of *Camera Obscura* 29: 131–50.

———. 1997. *Cultural Conceptions: On Reproductive Technologies and the Remaking of Life*. Minneapolis: University of Minnesota Press.

Hausman, Bernice L. 1995. *Changing Sex: Transsexualism, Technology, and the Idea of Gender*. Durham, N.C.: Duke University Press.

Hayles, Katherine. 1999. *How We Became Posthuman: Virtual Bodies in Cybernetics, Literature, and Informatics*. Chicago: University of Chicago Press.

———. 2000. "Flickering Connectivities in Shelley Jackson's *Patchwork Girl:* The Importance of Media-Specific Analysis." *Postmodern Culture* 10(2). Order online at http://muse.jhu.edu/journals/postmodern_culture/v010/10.2hayles.html (subscription required; link accessed Feb. 21, 2007).

———. 2002. *Writing Machines*. Cambridge, Mass.: MIT Press.

———. 2005. *My Mother Was a Computer: Digital Subjects and Literary Texts*. Chicago: University of Chicago Press.

Heim, Michael. 1991. "The Erotic Ontology of Cyberspace." In Michael Benedikt, ed., *Cyberspace: First Steps*. Cambridge, Mass.: MIT Press.

Hillis, Ken. 1999. *Digital Sensations, Space, Identity, and Embodiment in Virtual Reality*. Minneapolis: University of Minnesota Press.

Hissey, Arthur. Undated. "Your Life—on the Web." Online at http://www.crt.net.au/etopics/mylifebits.htm (retrieved March 17, 2007).

Hollinger, Veronica. 2002. "(Re)reading Queerly: Science Fiction, Feminism, and the Defamiliarization of Gender." In Mary Flanagan and Austin Booth, eds., *Reload: Rethinking Women + Cyberculture*. Cambridge, Mass.: MIT Press.

Homans, Margaret. 1986. *Bearing the Word: Language and Female Experience in Nineteenth-Century Women's Writing*. Chicago: University of Chicago Press.

Hoskins, Andrew. 2001. "New Memory: Mediating History." *Historical Journal of Film, Radio and Television* 21(4): 333–46.

Huyssen, Andreas. 1981. "The Vamp and the Machine: Technology and Sexuality in Fritz Lang's *Metropolis*." *New German Critique* 24/25: 221–37.

———. 1986. *After the Great Divide: Modernism, Mass Culture, Postmodernism*. Bloomington: Indiana University Press.

Irigaray, Luce. 1985. *Speculum of the Other Woman*. Trans. Gillian Gill. Ithaca, N.Y.: Cornell University Press.

Jackson, Shelley. 1995. *Patchwork Girl; Or, a Modern Monster*. Hypertext novel. Watertown, Mass.: Eastgate Systems. Order online at http://www.eastgate.com/catalog/PatchworkGirl.html (purchase required; link accessed Feb. 21, 2007).

Jacobus, Mary. 1982. "Is There a Woman in This Text?" *New Literary History, A Journal of Theory and Interpretation* 14(1): 117–41.

———, Evelyn Fox Keller, and Sally Shuttleworth. 1990. *Body/Politics: Women and the Discourses of Science*. New York: Routledge.

Johnson, Richard, et al. 2004. *The Practice of Cultural Studies*. Thousand Oaks, Calif.: Sage.

Jordanova, Ludmilla, ed. 1986. *The Languages of Nature: Critical Essays on Science and Literature*. London: Free Association Books.

———. 1989. *Sexual Visions: Images of Gender in Science and Medicine between the Eighteenth and Twentieth Centuries*. Madison: University of Wisconsin Press.

Joyce, Michael. 1997. "Nonce Upon Some

Times: Rereading Hypertext Fiction." *Modern Fiction Studies* 43(3): 579–97. Order online at http://muse.jhu.edu/journals/modern_fiction_studies (subscription required; link accessed Feb. 21, 2007).

Kaplan, E. Ann, ed. 1980. *Women in Film Noir*, rev. ed. London: British Film Institute.

———. 1998. *Women in Film Noir*, 2nd ed. London: British Film Institute.

Katriel, Tamar, and Thomas Farrell. 1991. "Scrapbooks as Cultural Texts: An American Art of Memory." *Text and Performance Quarterly* 11: 1–17.

Keller, Evelyn Fox. 1985. *Reflections on Gender and Science*. New Haven, Conn.: Yale University Press.

———. 1992. *Secrets of Life, Secrets of Death: Essays on Language, Gender and Science*. London: Routledge.

———. 1995. *Refiguring Life: Metaphors of Twentieth-Century Biology*. New York: Columbia University Press.

———. 1999. "The Gender/Science System? Or, Is Sex to Gender as Nature Is to Science?" In Mario Biagioli, ed., *The Science Studies Reader*. New York: Routledge.

———. 2000. *The Century of the Gene*. Cambridge, Mass.: Harvard University Press.

Kember, Sarah.2005. "Doing Technoscience as ('New') Media." In James Curran and David Morley, eds., *Media and Cultural Theory*. London: Routledge.

Kendrick, Michelle. 2001. "Interactive Technology and the Remediation of the Subject of Writing." *Configurations* 9: 231–51.

Kilborn, Richard, and John Izod. 1997. *An Introduction to Television Documentary: Confronting Reality*. Manchester: Manchester University Press.

Kirby, David. 2000. "The New Eugenics in Cinema: Genetic Determinism and Gene Therapy in *Gattaca*." *Science Fiction Studies* 27: 193–215.

Kirejczyk, Marta. 1994. "Cassandra's Warnings: Feminist Discourse, Gender and Social Entrenchment of In Vitro Fertilization in the Netherlands." *European Journal of Women's Studies* 2: 151–65.

———, Dymphie van Berkel, and Tsjalling Swierstra. 2001. *Nieuwe Voortplanting: Afscheid van de ooivaar*. Den Haag: Rathenau Instituut.

Kitzmann, Andreas. 2001. "Pioneer Spirits and the Lure of Technology: Vannevar Bush's Desk, Theodor Nelson's World." *Configurations* 9: 441–59.

Krimsky, Sheldon. 2000. *Hormonal Chaos: The Scientific and Social Origins of the Environmental Endocrine Hypothesis*. Baltimore, Md.: Johns Hopkins University Press.

Kristeva, Julia. 1984. *Revolution in Poetic Language*. New York: Columbia University Press.

Kuhn, Annette, ed. 1990. *Alien Zone: Cultural Theory and Contemporary Science Fiction Cinema*. London: Verso.

———, ed. 1999. *Alien Zone II: The Spaces of Science Fiction Cinema*. London: Verso.

———. 2000. "A Journey through Memory." In Susannah Radstone, ed., *Memory and Methodology*. Oxford: Berg.

Kuhn, Thomas. 1962. *The Structure of Scientific Revolutions*. Chicago: University of Chicago Press.

Kung, Hans. 1998. *A Global Ethic for Global Politics and Economics*. Oxford: University Press.

Kylhammar, Martin. 1985. *Maskin och idyll: Teknik och pastorala ideal hos Strindberg och Heidenstam*. Malmö, Sweden: Liber.

Landow, George P. 1997. *Hypertext 2.0: The Convergence of Contemporary Critical Theory and Technology*. Baltimore, Md.: Johns Hopkins University Press.

———, and Paul Delany, eds. 1993. *The Digi-*

tal Word: Text-Based Computing in the Humanities. Cambridge, Mass.: MIT Press.

Lanham, Richard. 1993. *The Electronic Word: Democracy, Technology, and the Arts*. Chicago: University of Chicago Press.

Leader, Zachary. 1996. *Revision and Romantic Authorship*. Oxford: Clarendon Press.

Lean, Geoffrey, and Richard Sadler. 2002a. "Male Fertility Fears over Pollution in Water Supply." *The Independent*, March 17.

———. 2002b. "British Men Are Less Fertile than Hamsters." *The Independent*, March 19.

Lefanu, Sarah. 1988. *In the Chinks of the World Machine: Feminism and Science Fiction*. London: Women's Press.

Levine, George. 2000. Foreword. In Gillian Beer, *Darwin's Plots: Evolutionary Narrative in Darwin, George Eliot, and Nineteenth-Century Fiction*, 2nd ed. New York: Cambridge University Press.

Lykke, Nina. 2002. "Feminist Cultural Studies of Technoscience and Other Cyborg Studies: A Cartography." In R. Braidotti, J. Nieboer, and S. Hirs, eds., *The Making of European Women's Studies*, vol. 4. Utrecht: ATHENA (Advanced Thematic Network in Activities in Women's Studies in Europe), Utrecht University.

———, and Rosi Braidotti, eds. 1996. *Between Monsters, Goddesses and Cyborgs: Feminist Confrontations with Science, Medicine and Cyberspace*. London: Zed Books.

———, and Mette Bryld. 2002. "Cyborg in Drag? Lennart Nilsson: Faszination Liebe: Der wissenschaftliche Dokumentarfilm zwischen Dekonstruktion und Naturalisierung." In M.-L. Angerer, K. Peters, and Z. Sofoulis, eds., *Future Bodies: Zur Visualisierung von Körpern in Science und Fiction*. Vienna: Springer-Verlag.

———, Randi Markussen, and Finn Olesen. 2000. "Cyborgs, Coyotes and Dogs: A Kinship of Feminist Figurations." Interview with Donna Haraway. In *Kvinder, Køn og Forskning* 2: 6–16.

———, Randi Markussen, and Finn Olesen. 2004. "Cyborgs, Coyotes and Dogs." Interview with Donna Haraway. In Donna Haraway, *The Haraway Reader*. London: Routledge.

Mallon, Thomas. 1984. *A Book of One's Own: People and Their Diaries*. New York: Ticknor and Fields.

Marcus, George. 1995. *Technoscientific Imaginaries: Conversations, Profiles, and Memoirs*. Chicago: University of Chicago Press.

Marshall, Tim. 1995. *Murdering to Dissect: Grave-robbing, Frankenstein and the Anatomy Literature*. Manchester: Manchester University Press.

Martin, Emily. 1991. "The Egg and the Sperm: How Science Has Constructed a Romance Based on Stereotypical Male-Female Roles." *Signs* 3: 485–501.

———. 1992. *The Woman in the Body: A Cultural Analysis of Reproduction*, 2nd. ed. Boston: Beacon Press.

———. 1994. *Flexible Bodies: Tracking Immunity in American Culture from the Days of Polio to the Age of AIDS*. Boston: Beacon Press.

Mayberry, Maralee, Banu Subramaniam, and Lisa H. Weasel, eds. 2001. *Feminist Science Studies: A New Generation*. New York: Routledge.

M'charek, Amade. 2005. *The Human Genome Diversity Project: An Ethnography of Scientific Practice*. Cambridge: Cambridge University Press.

McNeil, Maureen, and Sarah Franklin. 1991. "Science and Technology: Questions for Cultural Studies and Feminism." In Sarah Franklin, Celia Lury, and Jackie Stacey, eds., *Off-Centre: Feminism and Cultural Studies*. London: HarperCollins Academic.

McQuire, Scott. 1998. *Visions of Modernity: Representations, Memory, Time and Space in the Age of the Camera*. London: Sage.

Medvei, Victor Cornelius. 1982. *A History of Endocrinology*. Lancaster, England: MTP Press.

Menser, Michael, and Stanley Aronowitz. 1996. "A Manifesto on the Cultural Study of Science and Technology." In Stanley Aronowitz et al., eds., *Technoscience and Cyberculture*. New York: Routledge.

Minh-ha, Trinh T. 1986–87. "She, the Inappropriated Other." *Discourse* 8: 00–00.

Moers, Ellen. 1977. *Literary Women*. Garden City, N.Y.: Anchor Press.

Mol, Annemarie. 2003. *The Body Multiple: Ontology in Medical Practice*. Durham, N.C.: Duke University Press.

Mol, Annemarie, and John Law. 1994. "Regions, Networks and Fluids: Anaemia and Social Topology." *Social Studies of Science* 24: 641–71.

Moran, James M. 2002. *There's No Place Like Home Video*. Minneapolis: University of Minnesota Press.

Moulthrop, Stuart. 1991. "Reading from the Map: Metonymy and Metaphor in the Fiction of Forking Paths." In Paul Delany and George P. Landow, eds., *Hypermedia and Literary Studies*. Cambridge, Mass.: MIT Press.

———, and Nancy Kaplan. 1991. "Something to Imagine." *Computers and Composition* 9(1): 7–24.

Muensterberger, Werner. 1994. *Collecting: An Unruly Passion*. Princeton, N.J.: Princeton University Press.

Naess, Arne. 1977. "Spinoza and Ecology." In Siegfried Hessing, ed., *Speculum Spinozanum, 1877–1977*. London: Routledge & Kegan Paul.

National Research Council. 1999. *Hormonally Active Agents in the Environment*. Washington, D.C.: National Academy Press.

Nelkin, Dorothy. 1996. "Perspectives on the Evolution of Science Studies." In Stanley Aronowitz et al., eds., *Technoscience and Cyberculture*. New York: Routledge.

———, and M. Susan Lindee. 1995. *The DNA Mystique: The Gene as a Cultural Icon*. New York: Freeman.

Newbold, R. R., R. B. Hanson, W. N. Jefferson, B. C. Bullock, J. Haseman, and J. A. McLachlan. 1998. "Increased Tumors but Uncompromised Fertility in the Female Descendants of Mice Exposed Developmentally to Diethylstilbestrol." *Carcinogenesis* 19(9): 1655–63.

Newman, K., ed. 2002. *Science Fiction/Horror: A Sight and Sound Reader*. London: British Film Institute.

Nilsson, Lennart. 1965. "The Drama of Life before Birth." *Life*, Apr. 30: 54–72.

———. 1993. (Orig. 1966.) *A Child Is Born*. New York: DTP/Seymour Lawrence.

Nottingham, Stephen. 2000. *Screening DNA: Exploring the Cinema-Genetics Interface*. Stevenage: DNA Books.

Novas, Carlos, and Nikolas Rose. 2000. "Genetic Risk and the Birth of the Somatic Individual." *Economy and Society* 29(4): 485–513.

Oudshoorn, Nelly. 1994. *Beyond the Natural Body: An Archeology of Sex Hormones*. London: Routledge.

Paper Tiger Television. 1987. *Donna Haraway Reads 'The National Geographic' on Primates* (video no. 126). New York: Paper Tiger Television. Order online at http://www.papertiger.org.

Parens, Erik, and Erik Juengst. 2001. "Inadvertently Crossing the Germ Line." *Science* 292: 397.

Parisi, Luciana. 2004. *Abstract Sex: Philoso-*

phy, Bio-Technology, and the Mutation of Desire. London: Continuum Press.

Pasveer, Bernike, and Sara Heesterbeek. 2001. *De voortplanting verdeeld: De praktijk van de voortplantingsgeneeskunde doorgelicht vanuit het perspectief van patiënten*. The Hague: Rathenau Instituut.

Pearce, Lynne. 1994. *Reading Dialogics*. London: Edward Arnold.

Pearce, Susan. 1999. *On Collecting: An Investigation into Collecting in the European Tradition*. New York and London: Routledge.

Penley, Constance. 1997. *NASA/Trek: Popular Science and Sex in America*. London: Verso Books.

———, Elizabeth Lyon, Lynne Spigel, and Janet Bergstrom, eds. 1991. *Close Encounters: Film, Feminism, and Science Fiction Film*. Minneapolis: University of Minnesota Press.

———, and Andrew Ross, eds. 1991. *Technoculture*. Minneapolis: University of Minnesota Press.

Petchesky, Rosalind. 1987. "Fetal Images: The Power of Visual Culture in the Politics of Reproduction." In Michelle Stanworth, ed., *Reproductive Technologies: Gender, Motherhood and Medium*. Minneapolis: University of Minnesota Press.

Phillips, Adam. 1998. "Sameness Is All." In Martha C. Nussbaum and Cass R. Sunstein, eds., *Clones and Clones: Facts and Fantasies about Human Cloning*. New York: Norton.

Pickering, Andrew. 1992. *Science as Practice and Culture*. Chicago: University of Chicago Press.

Plant, Sadie. 1995. "The Future Looms: Weaving Women and Cybernetics." *Body & Society* 1(3–4): 45–64.

———. 1997. *Zeros and Ones: Digital Women and the New Technoculture*. New York: Doubleday.

Ploeg, Irma van der. 1998. "Prosthetic Bodies: Female Embodiment in Reproductive Technologies." Ph.D. diss., University of Maastricht.

Rabinow, Paul. 1992. "Artificiality and Enlightenment: From Sociobiology to Biosociality." In J. Crary and S. Kwinter, eds., *Incorporations*. New York: Zone Books.

———. 1996. *Making PCR: A Story of Biotechnology*. Chicago: University of Chicago Press.

Radway, Janice. 1998. "Gender in the Field of Ideological Production: Feminist Cultural Studies, the Problem of the Political Subject, and the Aims of Knowledge Production." In Sue Rosse et al., eds., *New Perspectives in Gender Studies: Research in the Fields of Economics, Culture and Life Sciences*. Stockholm: Almqvist & Wiksell.

Rapp, Rayna. 1999. *Testing Women, Testing the Fetus: The Social Impact of Amniocentesis in America*. New York: Routledge.

Reinel, Birgit. 1999. "Reflections on Cultural Studies of Technoscience." *European Journal of Cultural Studies* 2(2): 163–89.

"River 'Pollution' Sparks Fertility Fears." 2002. Online at http://news.bbc.co.uk/2/hi/uk_news/1877162.stm (retrieved Feb. 21, 2007).

Roberts, Celia. 2000. "Biological Behavior? Hormones, Psychology and Sex." *NWSA Journal* 12(3): 1–20.

———. 2002a. "A Matter of Embodied Fact: Sex Hormones and the History of Bodies." *Feminist Theory* 3(1): 7–26.

———. 2002b. "'Successful Aging' with Hormone Replacement Therapy: It May Be Sexist, but What If It Works?" *Science as Culture* 11(1): 39–59.

———. 2003a. "Drowning in a Sea of Estrogens: Sex Hormones, Sexual Reproduction and Sex." *Sexualities* 6(2): 195–213.

———. 2003b. "Sex, Race and 'Unnatural'

Difference: Tracking the Chiastic Logic of Menopause-Related Discourses." *European Journal of Women's Studies* 11(1): 27–44.

———. 2007. *Messengers of Sex: Hormones, Biomedicine and Feminism.* Cambridge: Cambridge University Press.

Roberts, Robin. 1993. *A New Species: Gender and Science in Science Fiction.* Urbana: University of Illinois Press.

Romney, Jonathan. 1998. "_______ ." *Sight and Sound,* March: 48–49.

Roof, Judith. 2007. *The Poetics of DNA.* Minneapolis: University of Minnesota Press.

Rose, Hilary. 1994. *Love, Power and Knowledge: Towards a Feminist Transformation of the Sciences.* Cambridge: Polity Press.

Rose, Nicholas. 2001. "The Politics of Life Itself." *Theory, Culture & Society* 18(6): 1–30.

Rosinsky, Natalie M. 1984. *Feminist Futures: Contemporary Women's Speculative Fiction.* Ann Arbor, Mich.: UMI Research Press.

Ross, Andrew. 1991. *Strange Weather: Culture, Science and Technology in the Age of Limits.* London: Verso.

Rouse, Joseph. 1992. "What Are Cultural Studies of Scientific Knowledge?" *Configurations* 1: 1–22.

The Royal Society. 2000. *Endocrine-Disrupting Chemicals (EDCs).* London: The Royal Society.

Rubenstein, D., et al. 1995. "Germ-Line Therapy to Cure Mitochondrial Disease: Protocol and Ethics of In Vitro Ovum Nuclear Transplantation." *Cambridge Quarterly of Healthcare Ethics* 4: 316–39.

Sawchuk, K. 2000. "Biotourism, Fantastic Voyage and Sublime Inner Space." In K. Sawchuk and J. Marchessault, eds., *Wild Science: Reading Feminism, Medicine and the Media.* London: Routledge.

Scheeres, Julia. 2002. "Saving Your Bits for Posterity." *Wired News* 6. Online at http://www.wired.com/news/technology/0,1282,56734,00.html (retrieved March 17, 2007).

Schneider, David M. 1968. *American Kinship: A Cultural Account.* Englewood Cliffs, N.J.: Prentice-Hall.

———. 1984. *A Critique of the Study of Kinship.* Ann Arbor: University of Michigan Press.

Schrödinger, Erwin. 1967. (Orig. 1944.) *What Is Life? The Physical Aspect of the Living Cell, & Mind and Matter.* Cambridge: Cambridge University Press.

———. 1983. (Orig. 1935.) "On the Present Situation in Quantum Mechanics." Trans. J. D. Trimmer. In J. A. Wheeler and W. H. Zurek, eds., *Quantum Theory and Measurement.* Princeton, N.J.: Princeton University Press.

Schwarz, Hillel. 1996. *The Culture of the Copy: Striking Likeness, Unreasonable Facsimiles.* New York: Zone Books.

Science and Technology Subgroup. 1991. "In the Wake of the Alton Bill: Science, Technology and Reproductive Politics." In Sarah Franklin, Celia Lury, and Jackie Stacey, eds., *Off-Centre: Feminism and Cultural Studies.* London: HarperCollins Academic.

Seaman, Barbara. 2003. *The Greatest Experiment Ever Performed on Women: Exploding the Estrogen Myth.* New York: Hyperion.

———, and Gideon Seaman. 1977. *Women and the Crisis in Sex Hormones.* New York: Bantam.

Sharpe, Richard M., and D. Stewart Irvine. 2004. "How Strong Is the Evidence of a Link between Environmental Chemicals and Adverse Effects on Human Reproductive Health?" *British Medical Journal* 328: 447–51.

Shelley, Mary. 1994. (Orig. 1818.) *Frankenstein: Or, the Modern Prometheus.* London: Penguin Books.

Shiva, Vandana. 1997. *Biopiracy: The Plunder of Nature and Knowledge*. Boston: South End Press.

Shohat, Ella. 1992. "'Laser for Ladies': Endo Discourse and the Inscription of Science." In Paula Treichler and Lisa Cartwright, eds., *Imaging Technologies, Inscribing Science*. Special issue of *Camera Obscura* 29: 57–89.

Silver, Lee M. 2000. "Reprogenetics: How Do a Scientist's Own Ethical Deliberations Enter into the Process?" Online at ________ (retrieved June __, 2007).

Silverman, Kaja. 1988. *The Acoustic Mirror: The Female Voice in Psychoanalysis and Cinema*. Bloomington: Indiana University Press.

Snow, C.P. 1993. (Orig. 1959.) *The Two Cultures*. Cambridge: Cambridge University Press.

Sobchack, Vivian. 1997. *Screening Space: The American Science Fiction Film*, 2nd ed. New Brunswick, N.J.: Rutgers University Press.

———. 2004. *Carnal Thoughts: Embodiment and Moving Image Culture*. Berkeley: University of California Press.

"Software Aims to Put Your Life on a Disk." 2002. *New Scientist*, Nov. 20. Online at http://www.newscientist.com/article.ns?id=dn3084 (retrieved March 17, 2002).

Solomon, Gina M., and Ted Schettler. 2000. "Endocrine Disruption and Potential Human Health Implications." *Canadian Medical Association Journal* 163(11): 1471–76.

Spanier, Bonnie B. 1995. *Im/partial Science: Gender Ideology in Molecular Biology*. Bloomington: Indiana University Press.

Spivak, Gayatri Chakravorty, with Ellen Rooney. 1994. "In a Word: Interview." In Naomi Schor and Elizabeth Weed, eds., *The Essential Difference*. Bloomington: Indiana University Press.

Springer, Claudia. 1996. *Electronic Eros: Bodies and Desire in the Postindustrial Age*. Austin: University of Texas Press.

———. 1999. "Psycho-Cybernetics in Films of the 1990s." In Annette Kuhn, ed., *Alien Zone II: The Spaces of Science Fiction Cinema*. London: Verso.

Stabile, Carol A. 1992. "Shooting the Mother: Fetal Photography and the Politics of Disappearance." In Paula Treichler and Lisa Cartwright, eds., *Imaging Technologies, Inscribing Science*. Special issue of *Camera Obscura* 28: 179–205.

Stacey, Jackie. 1997. *Teratologies: A Cultural Study of Cancer*. London: Routledge.

———. 2000. "The Global Within: Consuming Nature, Embodying Health." In Sarah Franklin, Celia Lury and Jackie Stacey, *Global Nature, Global Culture*. London: Sage.

———. 2003. "She Is Not Herself: The Deviant Relations of Alien Resurrection." *Screen* 44(3): 251–76.

———. 2005. "Masculinity, Masquerade, and Genetic Impersonation: *Gattaca*'s Queer Visions." *Signs* 30(3): 1851–79.

Stelarc. 2000. "From Psycho-Body to Cyber-Systems: Images as Post-Human Entities." In D. Bell and B. Kennedy, eds., *The Cybercultures Reader*. London: Routledge.

Storey, John. 1993. *An Introduction to Cultural Theory and Popular Culture*. Athens: University of Georgia Press.

Strathern, Marilyn. 1987. "Out of Context: The Persuasive Fictions of Anthropology." *Current Anthropology* 3: 251–81.

———. 1992a. *After Nature: English Kinship in the Late Twentieth Century*. Cambridge: Cambridge University Press.

———. 1992b. *Reproducing the Future: Essays on Anthropology, Kinship and the New Reproductive Technologies*. Manchester: Manchester University Press.

———. 2002. "Still Giving Nature a Helping Hand? Surrogacy: A Debate about Technology and Society." *Journal of Molecular Biology* 319: 985–93.

Sturken, Marita, and Lisa Cartwright. 2001. *Practices of Looking: An Introduction to Visual Culture*. Oxford: Oxford University Press.

Sundén, Jenny. 2003. *Material Virtualities: Approaching Online Textual Embodiment*. New York: Peter Lang.

———. 2006. "Digital Geographies: From Storyspace to Storied Places." In J. Falkheimer and A. Jansson, eds., *Geographies of Communication: The Spatial Turn in Media Studies*. Göteborg: Nordicom.

Telotte, J. P. 1999. *A Distant Technology: Science Fiction Film and the Machine Age*. Hanover, N.H.: University Press of New England.

———. 2001. *Science Fiction Film*. Cambridge: Cambridge University Press.

Terry, Jennifer. 1999. *An American Obsession: Science, Medicine, and Homosexuality in Modern Society*. Chicago: University of Chicago Press.

Thompson, John B. 1995. *The Media and Modernity: A Social Theory of the Media*. Cambridge: Polity Press.

Thornham, Sue. 2000. *Feminist Theory and Cultural Studies*. London: Arnold.

Treichler, Paula. 1999. *How to Have Theory in an Epidemic: Cultural Chronicles of AIDS*. Durham, N.C.: Duke University Press.

———, and Lisa Cartwright, eds. 1992a. *Imaging Technologies, Inscribing Science*. Special issue of *Camera Obscura* 28 (entire issue).

———, eds. 1992b. *Imaging Technologies, Inscribing Science*. Special issue of *Camera Obscura* 29 (entire issue).

———. 1992c. Introduction. In Paula Treichler and Lisa Cartwright, eds., *Imaging Technologies, Inscribing Science*. Special issue of *Camera Obscura* 28: 1–18.

———, and Constance Penley, eds. 1998. *The Visible Woman: Imaging Technologies, Gender, and Science*. New York: New York University Press.

Turner, Graeme. 2003. *British Cultural Studies*, 3rd ed. London: Sage.

Turney, Jon. 1998. *Frankenstein's Footsteps: Science, Genetics, and Popular Culture*. New Haven, Conn.: Yale University Press.

Urry, John. 2003. *Global Complexity*. Cambridge: Polity Press.

Veeder, William. 1986. *Mary Shelley and Frankenstein: The Fate of Androgyny*. Chicago: University of Chicago Press.

Virilio, Paul. 1998. *The Virilio Reader*. Ed. James Der Derian. Oxford: Blackwell.

Waldby, Catherine. 2002. "The Instruments of Life: Frankenstein and Cyberculture." In D. Tofts, A. Jonson, and A. Cavallaro, eds., *Prefiguring Cyberculture: An Intellectual History*. Cambridge, Mass.: MIT Press.

Waldén, Louise. 1990. *Genom symaskinens nålsöga: Teknik och social förändring i kvinnokultur och manskultur*. Stockholm: Carlssons.

Williams, Raymond. 1973. *The Country and the City*. London: Chatto and Windus.

———. 1974. *Television: Technology and Cultural Form*. London: Fontana.

———. 1980. "Ideas of Nature." First published in 1972; reprinted in his *Problems in Materialism and Culture: Selected Essays* (London: Verso).

———. 1988. *Keywords: A Vocabulary of Culture and Society*, rev. ed. Glasgow: Fontana.

———. 1989. *Towards 2000*. London: Chatto and Windus.

Wilson, E. O. 1975. *Sociobiology: New Synthesis*. Cambridge, Mass.: Belknap Press.

Winkler, Hartmut. 2001. "Discourses, Schemata, Technology, Monuments: Outline for a Theory of Cultural Continuity." *Configurations* 9: 91–109.

Women's Studies Group. 1978. *Women Take Issue: Aspects of Women's Subordination.* London: Hutchinson Centre for Contemporary Cultural Studies, University of Birmingham.

Wood, Aylish. 2002. *Technoscience in Contemporary American Film: Beyond Science Fiction.* Manchester: Manchester University Press.

World Wildlife Fund for Nature. 2002. "Who Cares Where Toxic Chemicals End Up?" *Country Living*, Feb.: 43, and *Observer Food Monthly*, Feb.: 25. Online at www.wwf.org.uk/Whocares/page3.asp (retrieved March 21, 2007).

Writing Group for the Women's Health Initiative Randomized Controlled Trial. 2002. "Risks and Benefits of Estrogen plus Progestin in Healthy Postmenopausal Women." *Journal of the American Medical Association* 288(3): 321–23.

Youngquist, Paul. 1991. "Frankenstein: The Mother, the Daughter, and the Monster." *Philological Quarterly* 70(3): 339–59.

Zaleski, Jeff. 1997. *The Soul of Cyberspace: How New Technology Is Changing Our Spiritual Lives.* New York: HarperCollins.

Zylinska, Joanna, ed. 2002. *The Cyborg Experiments: The Extensions of the Body in the Media Age.* London: Continuum.

Contributors

KAREN BARAD is professor of feminist studies, philosophy, and history of consciousness at the University of California, Santa Cruz. Her Ph.D. is in theoretical particle physics. She is the author of *Meeting the Universe Halfway: Quantum Physics and the Entanglement of Matter and Meaning* (2007) and of numerous articles on physics, philosophy of science, cultural studies of science, and feminist theory.

ROSI BRAIDOTTI is distinguished professor of the humanities in the arts faculty of Utrecht University. Until recently, she directed such European bodies as the Network of Interdisciplinary Women's Studies in Europe (NOI&SE, in the Erasmus program) and the Thematic Network for European Women's Studies (ATHENA, in the Socrates program). She has been a visiting professor at the Institute for Advanced Study, Princeton University; the European University Institute, in Florence; and the Law School of Birkbeck College, London. She has published widely on (French) philosophy and feminism. Her many books include *Transpositions: On Nomadic Ethics* (2006), *Metamorphoses: Towards a Materialistic Theory of Becoming* (2002), *Nomadic Subjects* (1994), and *Patterns of Dissonance: A Study of Women in Contemporary Philosophy* (1991).

METTE BRYLD has written extensively on literature and culture in various countries and has contributed to and edited a number of books, including *Soviet Civilization between Past and Present*, co-edited with Erik Kulavig (1998), and *Cosmodolphins: Feminist Cultural Studies of Technology, Animals, and the Sacred*, co-authored with Nina Lykke (2000). Until her retirement, in 2004, she was senior lecturer at the University of Southern Denmark.

JOSÉ VAN DIJCK is professor of media and culture at the University of Amsterdam. She is the author of *Manufacturing Babies and Public Consent: Debating the New Reproductive Technologies* (1995), *Imagenation: Popular Images of Genetics* (1998), *The Transparent Body: A Cultural Analysis of Medical Imaging* (2005), and *Mediated Memories in the Digital Age* (2007). Her research areas include media and science, (digital) media technologies, and television and culture.

GRIETJE KELLER is a graduate of the University of Amsterdam,where she studied the politics of breastfeeding. As a filmmaker, she has made several television documentaries. Her most recent film is about early computing and gender. Her work in progress includes a film for the International Information Center and Archives of the Women's Movement in Amsterdam, on oral histories of Dutch feminists.

NINA LYKKE is professor of gender studies, Linköping University, Sweden, and head of the Nordic Research School in Interdisciplinary Gender Studies (funded by the Nordic Council of Ministers funding agency NordForsk) and an international Center of Gender Excellence, GEXcel (funded by the Swedish Research Council). Her books include *Cosmodolphins: Feminist Cultural Studies of Technology, Animals, and the Sacred*, co-authored with Mette Bryld (2000), and *Between Monsters, Goddesses and Cyborgs: Feminist Confrontations with Science, Medicine and Cyberspace*, co-edited with Rosi Braidotti (1996).

RANDI MARKUSSEN, associate professor at the Institute of Information and Media Studies, Aarhus University, Denmark, is co-founder of the Centre for Science and Technology Studies at that institution. She has published numerous articles on science and technology studies.

AMADE M'CHAREK is associate professor in the Department of Political Science and in the Department of Biology at the University of Amsterdam. Her research focuses on the politics of diversity in medical and scientific practice and on issues concerning individuals, populations, sex, race, and related differences. She is the author of *The Human Genome Diversity Project: An Ethnography of Scientific Practice* (2005).

MAUREEN MCNEIL is professor of women's studies and cultural studies at Lancaster University, based in the Centre for Gender and Women's Studies, affiliated with the Centre for Science Studies, and a researcher in the Centre for Economic and Social Aspects of Genomics (CESAGen) there. Her forthcoming publications include *Human Cloning in the Media: From Science Fiction to Science Practice*, with Joan Haran, Jenny Kitzinger and Kate O'Riordan, and *Feminist Cultural Studies of Science and Technology*.

FINN OLESEN, associate professor at the Institute of Information and Media Studies, Aarhus University, Denmark, is co-founder of the Centre for Science and Technology Studies at that institution. He has published numerous articles on the philosophy of technology.

CELIA ROBERTS is a lecturer in the Department of Sociology, Lancaster University. She has worked on new reproductive technologies and genetic testing of embryos, on breast cancer, and on HIV/AIDS. She is the author of *Messengers of Sex: Hormones, Biomedicine and Feminism* (2007).

ANNEKE SMELIK is professor of visual culture and holds the Katrien van Munster chair at the Radboud University of Nijmegen (Netherlands). She is the author of *And the Mirror Cracked: Feminist Cinema and Film Theory* (1998) and co-editor, with Rosemarie Buikema, of *Women's Studies and Culture: A Feminist Introduction* (1995). She is also the principal author of the book *Effectief Beeldvormen* (1999), about representations of femininity and masculinity in the media, and co-editor of the book *Stof en As: Elf september in kunst en populaire cultuur* (2006), on art after September 11, 2001. Her research interests include digital art and culture, the performance of authenticity in fashion, and issues of multimedia literacy.

JACKIE STACEY was professor of women's studies and cultural studies in the Department of Sociology at Lancaster University until autumn 2007, when she took up a new chair in cultural studies at the Research Institute for Cosmopolitan Cultures at Manchester University. She is co-editor of the journals *Screen* and *Feminist Theory*. Her publications include *Star Gazing: Female Spectatorship and Hollywood Cinema* (1994), *Teratologies: A Cultural Study of Cancer* (1997), and *Global Nature, Global Culture*, co-authored with Sarah Franklin and Celia Lury (2000). She has completed a manuscript for Duke University Press titled *The Cinematic Life of the Gene* (forthcoming).

JENNY SUNDÉN is assistant professor of media technology in the School of Computer Science and Communication, Royal Institute of Technology (KTH), Stockholm. She received her Ph.D. in communication studies from Linköping University. She has published primarily on new media, cultural studies, cyberfeminism, virtual worlds, online ethnography, and digital textuality. She is the author of *Material Virtualities: Approaching Online Textual Embodiment* (2003) and co-editor, with Malin Sveningsson Elm, of *Cyberfeminism in Northern Lights: Digital Media and Gender in a Nordic Context* (2007).

Index

www.ingramcontent.com/pod-product-compliance
Ingram Content Group UK Ltd.
Pitfield, Milton Keynes, MK11 3LW, UK
UKHW041821090225
454784UK00002B/110

9 780295 997117